Adolph O. Pfingst, John E. Cashin

Manual of Elementary Bacteriology

Adolph O. Pfingst, John E. Cashin

Manual of Elementary Bacteriology

ISBN/EAN: 9783742830043

Manufactured in Europe, USA, Canada, Australia, Japa

Cover: Foto ©berggeist007 / pixelio.de

Manufactured and distributed by brebook publishing software
(www.brebook.com)

Adolph O. Pfingst, John E. Cashin

Manual of Elementary Bacteriology

MANUAL

OF

ELEMENTARY BACTERIOLOGY

BY

ADOLPH O. PFINGST, M. D.

Director Bacteriological Laboratory, Louisville Medical College; Ophthalmologist,
Aurist, and Laryngologist to City Hospital,
Louisville, Ky., Etc.

AND

JOHN E. CASHIN, M. D.

Demonstrator of Bacteriology in the Hospital College of Medicine and the Louisville
College of Dentistry, Louisville, Ky.; Bacteriologist State
Board of Health of Kentucky, Etc.

LOUISVILLE, KY.
JOHN P. MORTON AND COMPANY.
1898

PREFACE.

In presenting this manual we are endeavoring to meet a demand of the student for a briefer discussion of bacteriological principles than is found in the text-books now on the market.

It has been our object to give the present status of bacteriological knowledge in as concise a way as possible. To attain this we have avoided detailed description of old and obsolete methods which have only historic interest, and of theories which only tend to confound the student.

Attention has been paid to the practical needs of the student and physician by describing accurately the steps in the technique of bacteriological research, and entering fully into the fundamental principles of the subject.

Non-pathogenic bacteria have been omitted, to lend space to the consideration of the disease-producing varieties.

For reference we have made use of the works of Flügge, Guenther, Sternberg, McFarland, and Abbot.

A. O. P.
J. E. C.

INDEX.

PART FIRST.

GENERAL CONSIDERATION OF BACTERIA.

(a) INTRODUCTION. (b) MORPHOLOGY. (c) CONDITIONS OF GROWTH. (d) PRODUCTS OF VITAL ACTIVITY. (e) INFECTION. (f) IMMUNITY AND SUSCEPTIBILITY.

Part First.

GENERAL CONSIDERATION OF BACTERIA.

(a) INTRODUCTION.

Although bacteria were observed as early as 1675 by Leeuwenhoek, an ingenious Dutch optician of that time, the study did not develop into a science until the latter half of the present century. The intervening time represents a period of almost absolute quiescence. The pioneer work of Pasteur, Koch, and other authors of less prominence, and the introduction of new methods of investigation by them, including means of sterilization and of keeping culture media sterile, as well as the employment of suitable staining solutions (aniline dyes), lent much toward the establishment of the science.

However, the step of greatest importance was that introduced by Koch (1881) of employing solid culture media, thereby furnishing means wherewith the different species could readily be isolated. His method of isolation may be looked upon as the turning stone in bacteriological research. From the time of its introduction rapid strides have been made and many specific organisms discovered, so that there is hardly an enlightened practitioner of to-day who doubts that infectious disease is the product of germ activity in the body.

The benefits which the medical profession has derived from the new science have been felt principally in the regime of surgery and preventive medicine. It was the recognition of the fact that we are everywhere surrounded by bacteria, and the study of their character, growth, etc., that led to the introduction of aseptic surgery. Brilliant results in wound treatment and marked advances in surgical technique have been the result. This knowledge of bacteria also brought about the introduction of strict hygienic rules, which have been of inestimable value in the prevention of the spread of disease.

Bacteriological methods have, furthermore, become an invaluable means of diagnosis. It is true that in some diseases a bacteriological diagnosis can be made only by culture from the dejecta, blood, etc., which is too complicated to be of practical value in every-day work. However, in a number of diseases an immediate examination of the excretions, blood, etc., will reveal positive evidence of certain diseased conditions (pulmonary tuberculosis, gonorrhea, anthrax, etc.). The application to diagnosis of the late discovery, that the blood of individuals suffering with certain diseases would produce characteristic reaction with the pure culture of the germ causing the disease, has found favor and seems to be reliable. Its greatest practical benefit can, however, only be derived by those who have pure cultures always at their disposal.

The field from which the greatest achievements are to be expected is the employment of bacteria and their products as therapeutic measures. The serum-therapy, although in its infancy, has rapidly sprung into prominence. Its brilliant success in the treatment of diphtheria has sufficiently demonstrated the value of the principle involved, and will with certainty lead to the discovery of similar means of combating other diseases.

(b) MORPHOLOGY OF BACTERIA.

Bacteria (fission-fungi, schizomycetes,) are the lowest forms of vegetable life. They are very minute, unicellular organisms, consisting of a cell wall, more or less elastic, which incloses an apparently homogeneous or slightly granular transparent protoplasm. According to recent authors the protoplasm contains a nucleus which can be demonstrated by appropriate staining methods.

They are divided according to their shape into three great classes: (1) Bacillus, the filiform or rod-shaped cell; (2) Micrococcus, the spherical; (3) Spirillum, cells twisted or spiral. All bacteria may be classed into one of these three grand divisions. A cell under different conditions may vary in size and in form to a certain extent, but it is not possible by cultivation or other means to convert one species or class into another.

Size.—The micrococci vary in diameter from 0.15 μ to 2.8 μ; bacilli from 0.2 $\mu \times$ 0.4 μ to 4 $\mu \times$ 30 μ. The largest disease-producing bacillus, bacillus of anthrax, is about 1 $\mu \times$ 5 μ. The spiral forms have greater length, the longest being about 40 μ.

NOTE.—The unit of measurement of bacteria is represented by the Greek letter μ. 1 μ (micro-millimeter) is equal to $\frac{1}{1000}$ of a millimeter or $\frac{1}{25000}$ of an inch.

Chemical Composition.—Bacteria are composed chiefly of a peculiar albuminous substance, *mykoprotein*, the composition of which probably varies in different species, and in the same species under different conditions of growth. They also contain a small amount of fat. They do not contain starch or chlorophyll. There are other substances found in some varieties, among them sulphur and material peculiar to some species, by virtue of which they react in a specific manner to certain stains.

Capsule Bacteria.—Some of the bacteria are invested by a transparent, gelatinous substance, inclosing one or more cells. This investment, often spoken of as the *capsule*, is seen in bacteria, with few exceptions, only when growing within an animal body. It is thought to be merely an expansion of the cell wall. The capsule should not be confounded with the bright halo which, under the microscope, surrounds all unstained bacteria. The latter is a refractive phenomenon.

Zoöglœa.—An intercellular substance is sometimes seen binding together large numbers of bacteria in irregular masses called zoöglœa.

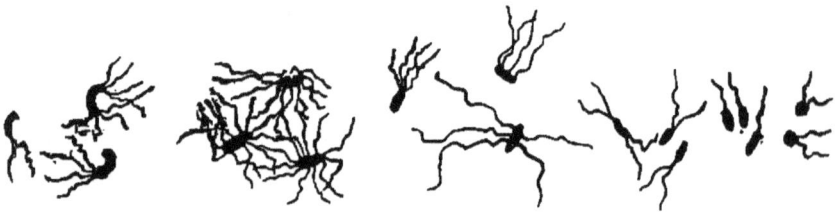

Fig. 1.

Motility.—Independent movement is observed in many species. It is influenced by temperature, age, vitality of the bacteria, etc. The motile bacteria are usually provided with delicate whip-like appendages, *cilia* or *flagella*. The number of cilia and their points of origin vary. (Fig. 1.) Sometimes a single flagellum can be seen projecting from one extremity of a cell; or both ends and sides may be supplied with them; sometimes a cell is provided with fifteen or twenty of these appendages. Cilia have not been found in all motile bacteria; and, as some non-motile species are richly provided with them, they are not regarded as essential to locomotion. It is probable that motility oftentimes is due to contraction and expansion of the elastic cell wall and the contents within. The micrococci, with but two or three exceptions, are non-motile and have no flagella. Special methods of staining are required for demonstrating these appendages.

Bacilli.—Under the name bacillus we describe all straight bacteria in which one diameter of the cell exceeds the other. They vary in thickness and length. Some are long, slender rods, others short and plump. The ends may be

Fig. 2.

pointed, blunt, rounded, or concave. (Fig. 2.) They occur singly or united in long chains, and also in zoöglœa masses. The term *leptothrix* is applied to long bacillary threads in which there is no apparent division into segments. The term *bacterium* was formerly applied to non-motile bacilli whose length was not greater than twice the breadth. This confounded a general with a specific term and has been abandoned.

Micrococci.—As micrococci— or simply cocci—we class all bacteria whose outline is spherical. They present wide differences in size and mode of grouping. If after division

Fig. 3.

the cocci remain united the adjacent surfaces of the young cells often appear flattened. In a few species the elements are lancet-shaped.

According to the mode of grouping, micrococci are subdivided into several varieties: (Fig. 3.)

(1) *Diplococcus*, in which the arrangement of cells is in pairs, united either by slightly pointed extremities or by adjacent flattened surfaces.

(2) *Tetragenococcus* (tetrad), in which division takes place in two directions, forming a group of four cells linked together.

(3) *Sarcinacoccus* (sarcina), in which division takes place in three directions, giving rise to regular, cubical, or bale-like packets.

(4) *Streptococcus*, in which division takes place in a straight line; the cells remain united, forming chains resembling strings of beads.

(5) *Staphylococcus*, in which the grouping is in irregular bunches, resembling clusters of grapes or the roe of fishes.

Spirilla are elongate organisms which have a curved or spiral (corkscrew-like) form.   The spiral form develops best in liquids.   Upon a solid medium the spirilla appear as rods more or less curved.   The term *vibrio* was applied by Ehrenberg to short, curved, flexible bacteria having

Fig. 4.

a sinuous motion.   At the present time the vibriones are classed with the spirilla.   (Fig. 4.)

Reproduction.—Multiplication in all forms of bacteria takes place by direct *cell division* (fission).   A cell about to undergo division is seen to enlarge and elongate somewhat; a constriction appears in the center of the long diameter of its wall; this gradually deepens until finally the cell is completely divided.   After division the young cells may remain united or separation may be complete.   The rapidity of this process varies in different species.   Under suitable conditions

of growth those which multiply most rapidly are capable of
undergoing fission in twenty minutes. The life of a single
generation of rapidly growing bacteria is from twenty minutes to one hour.

In the process of fission in the bacilli and spirilla division
occurs in a direction transverse to the long diameter of the
cell; in the cocci examples are found of division in one,
two, and three directions.

In addition to direct division, many of the rod-shaped and
a few spiral forms under certain conditions reproduce themselves, so to say, through spores formed within the cells—
sporulation. Spores are highly refractive, glistening bodies,
oval or rounded in form, which
develop either in the middle or at
one extremity of the cell. (Fig. 5.)
A cell bears but a single spore.
The process is in brief as follows:
Bright, glistening granules appear
in the cell contents which coalesce
into a round or oval body, the

Fig. 5.

spore. This is surrounded by a tough membrane, the exosporium or spore membrane. That part of the cell which
does not enter into the formation of the spore remains
perfectly clear, and after a time disappears, leaving a free
spore.

Sometimes a large central spore is formed which gives to
the cell a more or less spindle shape. Such a cell is spoken
of as a *clostridium.*

Brought again under suitable conditions, the spore will
develop into a cell identical with the one in which it was
formed. A spore about to germinate is observed to lose its
glistening appearance and increase in size. A protuberance
appears, enveloped by the inner lining, endosporium, which,
gradually increasing in size, ruptures the exosporium and
the young cell escapes.

Spore formation is most abundant under conditions highly favorable to growth. We can deprive some species of this property by cultivation under conditions inimical to development.

Involution Forms.—The vigor of development influences to a certain extent the form of bacteria. Under unfavorable conditions or in bacteria possessing feeble vitality variously shaped cells in nowise resembling the perfect organism are formed. These are spoken of as involution forms. (Fig. 6.) If sufficient vitality remains when conditions again become favorable they return to the normal type. Involution forms are produced quickly in highly virulent disease-producing bacteria. Those which have grown for a long time outside the body do not produce them so readily.

Fig. 6.

(c) CONDITIONS OF GROWTH.

Bacteria require for their development suitable food, warmth, and moisture. These requirements vary widely both as regards the amount of nutritive elements and the relative proportion in which each must be present. Some species will grow only in concentrated albuminous fluids, such as the blood-serum of animals. Others find in the small quantity of organic matter and salts present in water sufficient material to meet their requirements. Some obtain nutriment by oxidation of ammonia. A small number of species are able to obtain nutriment by oxidation of ethyl alcohol (vinegar bacterium), others from oxidation of H_2S. The artificial culture media for general use contain about eighty per cent of water, some salts, and about two per cent of organic matter. Nutritious material must enter the cell by osmosis; it is evident, therefore, that only watery solutions of diffusible material can be utilized; solid nutriment to be used must undergo liquefaction and decomposition before the required elements can be appropriated.

Food solutions, to be suitable to the largest number of species, should have a neutral or slightly alkaline reaction. A few require an acid medium.

By far the greater number of bacteria, finding in water, soil, dead animal and vegetable matter suitable material for nutrition, lead a *saprophytic* mode of life. A few species are strict *parasites*, growing only in the bodies of animals; still others, while finding external conditions favorable, sometimes invade the body; these are *facultative* or *occasional parasites*.

Carbon.—As the bacteria do not contain chlorophyll, the carbon present in the air as CO_2 can not be utilized by them. This element must be obtained by decomposing the tissues of plants and animals, and from suitable organic compounds, as glycerin, sugar, the fatty acids, and salts of vegetable acids.

Nitrogen is obtained from albuminoid substances, and to a limited extent from ammonia salts.

There is a class of bacteria, the *nitrifying organisms*, which can not appropriate the ordinary albuminous fluids used as culture media. They are cultivated in solutions containing certain mineral salts.

Oxygen.—Unrestricted access of air is indispensable to the full development of many species. In contrast with these are others which will not develop if oxygen is present. There are numerous forms which grow with or without this gas.

This difference in relation to oxygen has been made the basis of a biological division into the following groups: (1) *aerobic*, those growing only in the presence of oxygen; (2) *anaerobic*, those which do not grow in the presence of oxygen; (3) *facultative anaerobic*, those which, growing best in the presence of oxygen, are capable of development when it is limited or excluded.

It is obvious that a bacterium which penetrates and multiplies within the body of a living animal must be facultative anaerobic. Strictly anaerobic bacteria have been induced to grow in the presence of oxygen by gradually habituating them to the presence of the gas, and also by combining them in culture with other varieties.

*****Temperature.**—The temperature required for the development of bacteria differs for the different species. The strictly saprophytic forms find the heat of summer weather in temperate climates or the average warmth of living rooms most suitable. They can not endure the temperature of the living body, and hence can not become parasitic. The facultative parasites are capable of development within a wider range of temperature, multiplying either at room temperature

*****For scientific purposes the centigrade or Celsius thermometer is used.

or at blood heat. The strict parasites develop only at the temperature of the body (35° to 40° C.).

It may be said, in general terms, that bacteria develop between 10° and 40° C., very few, however, having this entire range. There is a minimum and maximum above and below which growth does not take place, and within this an optimum at which it is most abundant. Growth at excessive temperature was noted by Miguel and other observers, who have called attention to species, living in thermal waters, which require a temperature of 60° to 70° C.

The effect of low temperature depends upon the duration of exposure of the species and the vitality of the germ used for experimentation. Freezing inhibits the growth but does not destroy most varieties of bacteria. Repeated freezing and thawing is more prejudicial than continuous cold.

Bacteria not containing spores are killed outright by a temperature of 50° to 60° C. for ten minutes in the presence of moisture. This applies to all except the thermophylic species referred to above.

Moisture.—The presence of moisture is essential to the growth of all species of bacteria. They differ widely in this respect, some prospering only in watery fluids, while others are able to multiply on dried crackers and other foods containing the minutest quantity of water.

The influence of the absence of moisture is exemplified in the arid regions of the West. Flesh cut into strips may be exposed to the air for days without undergoing decomposition. The dry air quickly absorbs the moisture, thus making a condition inimical to bacterial development.

Light.—Ordinary diffuse daylight has no influence upon bacteria. On the contrary, color production, a common phenomenon, does not take place so well in the dark. Exposure to direct rays of the sun, the intense glare of the arc light, and, to a lesser extent, the incandescent light, have a destructive effect. The blue and violet rays of the spectrum are most unfavorable; the red rays almost inert. Access of oxygen is necessary to destroy bacteria by sunlight, but temperature of the air is unimportant.

Electricity.—The effect of electricity, using an induction current from a dynamo machine, has recently been studied and found to destroy vitality.

Exposure to the cathode (X) rays does not influence development.

Agitation. — Slow-continued motion does not influence growth, but prolonged violent agitation, as in the passage of water over rapids, causes death by molecular disintegration of the cells.

Association of Species.—In most cases we notice an antagonistic action between species growing together. This may be due to the appropriation of certain elements by those which grow most rapidly, or the products of growth may render the medium unsuitable. Less frequently the association of two species is favorable to the growth of each. That species which finds the environment most favorable ultimately takes the precedence.

(d) PRODUCTS OF VITAL ACTIVITY.

In the life processes of bacteria we have resulting a variety of phenomena, some of which play a very important part in the economy of nature.

1. Color Production. — Chromogenesis is one of the most common and striking of these phenomena. The chemical nature of bacterial pigments is not understood. It is not known whether they are secreted as such or result from a reaction between bacterial products and substances in the culture medium. If the cultivation is made upon the surface of a solid medium, the pigment in some species remains closely associated with the growth; in others it is soluble and diffuses through the medium. Color production can be temporarily destroyed by cultivation under unfavorable environments. It is restored upon a return to normal surroundings. It is worthy of note that pigments, with one exception (that of spirillum rubrum), are produced only in the presence of oxygen. All the colors of the spectrum are formed. Some pigments are fluorescent.

2. Phosphorescence.—The property of phosphorescence is found in several marine species. It is often observed upon the surface of salt water, upon dead fish, etc. We have no definite knowledge of the phosphorescent material or the mode of its production. An alkaline medium containing a certain proportion of sodium chloride is requisite.

3. Production of Heat.—In hay, malt, cotton, etc., which have been moistened and compressed, high heat is sometimes produced by bacterial development. It may be sufficient to lead to spontaneous combustion.

4. Production of Enzymes.—Little is known of the chemistry of these bodies, which are developed during the growth of vegetable cells. Their most notable property is, that a small quantity of the enzymes will decompose a large amount of material, they themselves undergoing no change. The enzymes can be separated from the living cells which produce them, and in this isolated state continue to bring about the same changes. They are non-dialyzable and are destroyed by a temperature of 60° C. Of these enzymes several different varieties have been described:

(a) *Diastatic enzymes*, which have the property of converting starch into the different sugars.

(b) *Peptonizing enzymes*, which transform albuminoid bodies into diffusible products. The liquefaction of gelatin, blood-serum, and other media containing albumin is due to the action of this class of ferments.

(c) *Coagulating enzymes*, which bring about coagulation of the casein in milk.

(d) *Urinary enzymes.* These convert urea and urates of the urine into ammonium carbonate, and hippuric acid into glycocoll and benzoic acid.

5. Fermentation.—Under certain conditions the growth of bacteria is followed by extensive changes in the nutrient material, with the production of various new compounds. To this process of decomposition the term fermentation is applied.

The process of fermentation is divided according to the nature of the changes taking place into (1) fermentation by decomposition, (2) fermentation by oxidation, (3) a combination of these—putrefaction.

(1) **Examples of fermentation by decomposition** are: (a) the alcoholic fermentation of sugar by means of the yeast plant; (b) the production of lactic and butyric acids in the fermentation of milk, cane sugar, starch, dextrin, etc.; (c) viscous fermentation, which sometimes takes place in sugar solutions; (d) the decomposition of cellulose.

Fermentation of Sugars.—Many bacteria have the power of decomposing sugar solutions with the evolution of CO_2, and an inflammable gas containing hydrogen. For testing this property lactose, glucose, or saccharose is added to bouillon in an Einhorn fermentation tube. Gas, if produced, collects at the top.

Fermentation of higher alcohols, fatty acids, and oxyacids is similar to that in the carbohydrates.

(2) **Fermentation by Oxidation.**— Examples of this fermentation are found in the production of acetic acid (vinegar) from weak solutions of alcohol and in nitrification of the soil.

(3) **Combined Fermentation — Putrefaction.**—This is a rapid and intense decomposition of albuminous substances, induced by a very large number of bacteria, during which foul-smelling gases are given off in large quantity. According to Pasteur putrefaction is largely the result of the multiplication of anaerobic species; certainly much of the odor of putrefaction is caused by them. The products of putrefactive fermentation vary widely, depending upon the composition of the material and the kinds of bacteria present. Among the substances are carbon dioxide, hydrogen, nitrogen, sulphuretted hydrogen, phosphoretted hydrogen, methane, and many volatile bodies, formic, acetic, butyric, oxalic, and many other acids, indol, skatol, and ptomaines.

6. Production of Indol, as a result of albuminoid decomposition, is of value in the determination of species. To detect its presence 1 c.cm. of one fiftieth per cent aqueous solution of potassium nitrite is added to 10 c.cm. of a bouillon culture; then one or two drops of concentrated sulphuric acid. If indol is present, a red color is produced. Among the pathogenic species which produce it are Bacillus coli communis and the spirillum of Asiatic cholera.

There are certain other complicated fermentations, whose chemistry is not well understood, which are used commercially and in the manufacture of articles of food. Among these are the fermentation in the manufacture of kumyss; cheese fermentation; the fermentation of bread, leather, tobacco, etc.

7. Ptomaines, Toxins, and Toxalbumins.— Ptomaines are basic nitrogenous compounds resulting from bacterial action upon organic matter, animal or vegetable. On account of their basic properties, in which they resemble vegetable alkaloids, they have been spoken of as animal or cadaveric alkaloids, but, inasmuch as they are produced also in vegetable matter, these terms are not appropriate. Some of these bodies are dangerous poisons (*toxic ptomaines or toxins*), others are wholly inert. Certain bacteria produce specific ptomaines, upon the absorption of which peculiar manifestations of disease depend. Such, for example, is tetanin, produced by the tetanus bacillus, which causes the muscular spasms of that disease. Besides specific ptomaines, there are a number of others, toxic and non-toxic, produced during putrefaction by the activity of any one of a number of different bacterial species.*

Those complex albuminous bodies formed during the growth of disease-producing bacteria in the animal economy and in artificial culture media are called *toxalbumins.* They are poisonous amorphous substances, soluble or insoluble in water. They are destroyed by a temperature of 60° C.

Closely allied to this class of proteid substances are the venoms of serpents and batrachia. Certain poisonous vegetable compounds are included among the toxalbumins.

The production of ptomaines and toxalbumins may be accounted for in two ways: Either they are decomposition products in albuminous bodies, or they are produced in the protoplasm of the bacterial cell and are excreted therefrom. As a modification of the first theory we may mention that which supposes the toxalbumins to be produced by the action of an enzyme or ferment upon the albuminous compounds of

---

*A leucomaine is a basic nitrogenous substance resulting from retrograde metamorphosis in living tissue.

the body.   In a certain degree the formation of toxins and toxalbumins by bacteria depends upon the composition of their nutriment and the virulence of the germ.

8. Production of Disease.— For the physician the most important action of bacteria is that of disease production. For its consideration see next chapter.

(e) INFECTION.

An important classification of bacteria is based upon their ability to give rise to disease by invasion of the animal body. Those having this property are spoken of as *pathogenic* or *infectious* bacteria, and the diseases caused by their proliferation *infectious diseases.* Parasitic and facultative parasitic bacteria are pathogenic, the saprophytic forms non-pathogenic. Some of the latter, however, generate poisonous ptomaines outside the body, the ingestion and absorption of which may produce disease.

The mere presence of bacteria in the lesions of a disease does not of itself signify that they have a causative relation thereto. In order that this may be established, certain other conditions must be fulfilled. The bacterium must be constantly present in the lesions of a particular disease, and in no other. It must cause the same pathological condition when introduced into another animal of the same species.

The Distribution of Bacteria in the Body is not the same in all infectious diseases, nor in all cases of the same disease. Certain bacteria seem to have a predilection for certain organs. Much depends, also, upon the virulence of the germ, the susceptibility of the animal, the mode and seat of infection. Infectious micro-organisms *may remain localized* at the point of entrance, as in tetanus and diphtheria, inducing disease by absorption of toxic products. Having obtained a foothold at some point, they *may gradually extend by continuity* of tissue. This is seen in erysipelas, in which the micro-organism invades the lymph-vessels of the skin.

In some diseases the field of activity is *limited to a certain organ,* as in pneumonia and tuberculosis. Again, we find in some diseases the seat of bacterial multiplication in the blood-vessels. This condition we speak of as *septicemia*. A bacterium which ordinarily induces a local disease may sometimes become generalized. A circumscribed process may be induced in one class of animals and a general infection (septicemia) in another by the same germ.

That form of general infection characterized by the formation of multiple abscesses throughout the body is designated *pyemia*. It differs from septicemia in the formation of capillary emboli composed of bacteria, with resulting abscesses.

Mixed infection is a term applied to the simultaneous entrance of two or more species of bacteria through the same channel. This type of infection may increase or decrease the severity and extent of the pathologic process; usually the former. It is quite distinct from *secondary infection*, in which a second pathologic process is engrafted upon the primary one.

In the course of various acute and chronic diseases, owing to the lowered vital powers, pathogenic bacteria normally present in or upon the body are able to invade the tissues and often are the immediate cause of death. The phrase *terminal infection* is applied to these cases.

Conditions Influencing Infection.—There are various factors which influence the development of pathogenic bacteria in the body. Many of these, particularly such as relate to the condition of the patient, are imperfectly understood. There can be no doubt that an enfeebled body-state and also the changes in the tissues induced by morbid states predispose to certain diseases. It is a matter of clinical experience that persons with diabetes are prone to tuberculosis and carbuncle. Much depends upon the *amount and concentration of the virus* and its malignancy. The larger the dose and the greater the virulence the more certain is infection to follow.

The *disease-producing power* of bacteria is their most variable characteristic. It *can be artificially diminished* (attenuated) by : Cultivation at high temperature ; exposure to sunlight; desiccation ; electricity ; increased atmospheric pressure ; exclusion of oxygen; long-continued growth upon artificial media; association in culture with other micro-organisms; growth in the body of an insusceptible animal; addition of blood-serum of immune animals; culture in extracts of thymus or lymphatic glands; the addition of germicidal agents to culture media.

Virulence may be intensified by: Slight increase of atmospheric pressure; transplantation at short intervals; growth in media containing glucose, glycerin, lactic acid, etc.; combination in culture with certain other micro-organisms; repeated inoculations into susceptible animals.

Pathology of Infection.—Infection, according to the modern acceptation of the term, is associated with the multiplication of micro-organisms in the tissues. It may be either local or general. It is customary to speak of diseases arising from the absorption of toxic products generated outside the body as *intoxications.* The term *toxic infection* is applied to those conditions resulting from the absorption of toxic products from a localized point of infection.

The proliferation of bacteria in the tissues is accompanied by constitutional symptoms and evidences of inflammatory changes at the point of infection. Sometimes we have a serous exudate; in some cases it is purulent, fibrinous, or hemorrhagic. There are also specific proliferative inflammations, as tuberculosis, leprosy, glanders, etc.

In addition to these local signs there are general evidences of poisoning due to absorption of toxic products. The symptom complex which we speak of as fever is always present in local infections. In septicemia it is inconstant. An increase in the number of leucocytes in the blood also accompanies local infections. There may be a diminished amount of hemoglobin with a tendency to hemorrhages. There are local nutritive disturbances as parenchymatous degeneration of the kidneys and muscle fibers, degenerative changes in the peripheral and central nervous systems, etc.

There are Various Avenues by which Pathogenic Germs gain Entrance to the System. They usually have a predilection for some particular portal.

(a) *By the Skin.*—With many bacteria it is possible to induce infection by rubbing them into the unbroken skin. As a rule, however, infection takes place through an abrasion or wound. Thus we find carbuncle occurring most commonly on the back of the neck. This may be explained by the friction of the collar, which destroys to some extent the epithelial layer of the cuticle. Recent breaches in the continuity of the skin and mucous membranes are most liable to infection. Suppurating and granulating surfaces as a rule do not absorb bacteria.

(b) *By the Mucous Membranes.*—Undoubtedly certain bacteria have the power of penetrating healthy mucous membranes (anthrax, diphtheria, gonorrhea). The poisonous ptomaines are rapidly absorbed, even though the bacteria remain localized.

(c) *The lung* offers a certain amount of resistance to the entrance of bacteria, and the air cells in a healthy state are free from them. The ciliated epithelium lining the air-passages is probably of some service in preventing their entrance. Aspiration of pathogenic bacteria contained in the oral and pharyngeal cavities is not uncommon during anesthesia; it also happens in diphtheria and other diseases of children, and in paralytic adults.

(d) *Stomach and Intestines.*—The germicidal power of acid gastric juice is to a high degree protective to the mucous membrane of the alimentary canal against infection. However, we often find bacteria, ordinarily killed by it, passing into the intestine unharmed. Examples of infection by this route are anthrax, Asiatic cholera, hog cholera, typhoid fever, etc.

(e) *Intra-uterine Infection.*—The placenta is not an impassable barrier to pathogenic bacteria. Infection of the child in utero occurs in anthrax, tuberculosis, syphilis, smallpox, and possibly other diseases.

(f) IMMUNITY AND SUSCEPTIBILITY.

Immunity is the ability of the animal body to prevent the entrance and development of pathogenic bacteria, or, when they have entered and developed, to withstand the action of their toxic products. We may accordingly speak of immunity to bacteria themselves or of immunity to their toxic products.

Susceptibility, the opposite of immunity, is a condition of the body in which little or no resistance is offered to the entrance of germs, their growth, and the production of disease.

Immunity may be *natural, acquired,* or *artificial.*

Natural Immunity.—We speak of natural immunity when the insuseptibility is inherent in the body at birth. Examples of it are familiar to every one in the immunity inherent in animals for diseases peculiar to man, viz., typhoid fever, yellow fever, Asiatic cholera, syphilis, etc.; also the immunity of man to diseases peculiar to certain species of lower animals. Certain races have a natural immunity to some diseases, the Negro being less susceptible to yellow fever and the Japanese to scarlet fever than the Caucasian.

Acquired Immunity.—The specific bacteria of certain infectious diseases bring about, during their activity in the body, a condition which makes it exempt from future attacks of the disease produced by them, or at least makes the second attack unlikely. A specific immunity is, in other words, acquired to the action of these certain germs and to the action of their toxic products. Immunity is acquired (a) *by recovery from a natural infection.* It is the general belief now that a specific immunity follows the recovery from all infectious diseases, but that it extends over a variable period of time. It is permanent after the exanthematous fevers, typhoid fever, etc. After an attack of pneumonia or erysipelas a very brief period of immunity follows. Later the individuals show a predisposition to the return of these diseases.

(*b*) *By recovery from a mild attack of a disease* resulting from inoculation of the disease-producing virus through other than the usual avenue of infection. The best example of this form of acquired immunity is that following the mild form of variola after vaccination.

Artificial Immunity can be brought about (*a*) *by the injection of the living germs specific for the disease.* As the virus would produce death in large doses, small quantities must be used. The same result can be brought about by using an attenuated virus, or, if a virulent culture be used, by injecting simultaneously with it chemical substances which modify its action (antiseptic solutions, normal blood-serum, etc.). The quantity and virulence can gradually be increased until the body is able to resist much larger quantities than would at the first injection have caused death. Immunity thus produced is almost purely a bacterial immunity. (*b*) *By the injection of products of cell growth.* They can be injected along with dead bacteria, in which case the living cultures are subjected to sufficient temperature (60° C. or more) to destroy life, or they are separated from the bacteria by filtration through porcelain. Beginning with a small quantity, it is gradually increased until large quantities are borne. The chemical composition of these products of germ life or their manner of bringing about immunity is not known. It protects against the toxin to a much greater extent than against the germ. (*c*) *By injecting tissue of immune animals, or extracts made from it (serum),* infection can be prevented in animals ordinarily susceptible. · This, in contradistinction to the former *active methods,* is referred to as the *passive method* of immunization. It has been shown that this germicidal or immunizing property exists, to some extent, in normal blood of immune as well as susceptible animals.

Influences Modifying Immunity.—Although we speak of immunity as being *relative* or *absolute*, the term is, strictly speaking, only a relative one, being modified by the age of the individual, climatic and hygienic surroundings, and various other influences brought to bear upon the body. Cultures attenuated so as to be almost harmless to a grown animal readily infect and kill young of the same species. It is well known that children and young animals are more highly susceptible to infectious diseases than the adult, and that animals under unfavorable climatic and hygienic surroundings are less able to resist infection than those more favorably situated.

Immunity sufficient to protect against disease under ordinary circumstances can be destroyed by methods which vary somewhat in each disease. The principle of all of them is to lower the vital resistant powers of the body.

Pasteur has shown that fowls refractory to anthrax can be infected by submersing them in cold water. Certain cold-blooded animals, on the other hand, may be infected by keeping them in water having a temperature of 35° C. White rats perish after subcutaneous injection or feeding with anthrax bacilli if previously fed with acid phosphate of sodium, which tends to lessen the alkalinity of the blood. Infection may be accomplished also by previously submitting the animal to exhausting labor, as was done in rats by confining them in a rotating cage. House-mice, which are naturally immune to glanders, may be infected by feeding them with phloridzin, which in these animals impregnates the tissues with sugar. Infection will take place in a diseased or injured organ which in a state of health would resist. The administration of poisons, such as chloral, curare, alcohol, etc.; the toxic products of the same, or of a different, species of bacterium, or a combination of different bacteria, are capable of destroying the natural immunity. Removal of the spleen acts in a similar manner.

Theories of Immunity.—At present all efforts to explain the difference of bodies in their resistance to disease are largely conjectural.

(*a*) **Natural (inherited) Immunity** has been ascribed to the inability of the body to furnish the proper or sufficient nutriment to certain germs; to unfavorable body temperature; unfavorable reaction of its juices, etc.

The most modern theory, and the one looked upon with most favor at present, is that advanced by Buchner. It is based upon the discovery that the fresh blood of various animals possesses germicidal properties. Certain proteids are supposed to pre-exist in the body, or to be produced by the leucocytes within it, which are antagonistic to certain bacterial action. Buchner called these substances *alexines*. They exist in the blood-serum, and can be separated by the addition of sulphate of sodium. Their chemical nature is unknown. They are supposed to be albuminoid bodies similar to nuclein. Alexines can exist in the body simultaneously with the toxins, which shows that their action is not dependent upon the destruction of germ poisons. Freezing and thawing have no effect upon them. Heating for one-half to one hour at 55° to 60° C. destroys them.

The theory of phagocytosis (Metschnikoff), which was the most popular several years ago, is now looked upon as of secondary importance. The supposition was that certain bacteria give off substances whose presence in the immune body attracts the movable body cell or leucocyte (positive chemiotaxis), that the leucocytes (phagocytes) pick up the germs and either destroy them within themselves or carry them to the spleen or lymphatic glands to be destroyed. In the susceptible body the chemiotaxic force repels the leucocytes, and consequently favors disease production. Some have attached to positive chemiotaxis importance in that it brings the leucocytes to the field of action, but believe that their real virtue consists in the liberation of alexines, which weaken or destroy the bacterial cells, thereby enabling the phagocyte to pick them up. This view is rendered probable by the observation that the more malignant the infection the rarer is the presence of bacteria within the phagocyte.

Notwithstanding the protective property of the body there are numerous virulent bacteria which are able to invade the organism and cause disease. The supposition is that these bacteria give off a substance which has an aggressive action for the alexine with which it unites in the formation of an inert body. This active substance has been called *lysine*. It combines with the alexines of the tissue, thereby rendering them powerless to antagonize the poisonous action of bacteria and their products.

(*b*) **Acquired and Artificial Immunity.**—Generally speaking, animals which have either artificial or acquired immunity resist only that particular germ with which the body had been infected. Sometimes a short period of immunity for other than the specific germ exists.

Many theories have been advanced to explain the acquisition of immunity by a previously susceptible body. Some of them have since been abandoned. Of the most plausible may be mentioned:

Metschnikoff's Theory.—This author has also applied his theory of phagocytosis to the explanation of specific immunity. As in natural immunity positive chemiotaxis is of undoubted importance, primarily, however, to bring the leucocytes and thereby the alexines; secondarily, to pick up weakened and dead bacteria.

Kruse's theory is that the serum of an immunized animal exerts a neutralizing influence on the lysine as rapidly as it is produced. Robbed of their lysine, the infectious bacteria can not develop on account of the combating influence of the alexines. They are like non-virulent germs. The affinity of the serum for lysine is believed by the author to be due to

a third substance which he called antilysine, and which is
supposed to be the result of union of lysine with elements of
the body cells.   In active immunity some of this substance
is stored in the body cells.   Passive (serum) immunity lasts
only as long as the antilysine is circulating in the blood.
New protective elements are not formed in the body made
immune by the passive method.

While this theory explains the immunity to bacterial
growth, it does not explain the specific immunity to their
poisons introduced into the body or produced within it.

The theory of antitoxins, which has been advanced to explain
this specific immunity to bacterial poisons, is the only one of
all of the hypotheses which is in any way substantiated by
experimental results.   (Immunization of Tetanus and Diphtheria by Kitasato and Behring.)   The teaching of this
theory is that a protective substance, antitoxin, is produced
as a result of a reactive tissue change taking place between
poisonous bacterial products and the tissue elements.   It is
consequently a new substance.   It in no way retards bacterial
growth, but prevents their toxic action by annulling the
toxin.   It is not destroyed as readily by heat as alexines and
antilysine, being able to resist 75° C.

Protective Inoculation and Serum-Therapy.—The
first instance of protective inoculation with a modified virus
was when Jenner, in 1776, introduced the contents of a
vesicle on the udder of a cow affected with cow-pox into
human beings as a protection against variola.   This vaccination, the introduction of living germs, whose virulence was
modified and fixed by passage through the slightly susceptible
cow, was followed by immunity.   No advances were made
along this line until the time of Pasteur, who made use of
living specific bacteria which had been modified in virulence
and "fixed" by artificial cultivation in the laboratory.
Having accidentally discovered that if the bacillus of fowl
cholera, which had been attenuated by cultivation for some

time, be inoculated in susceptible animals it was followed by immunity to the more virulent germ, he applied this principle to the prevention of several infectious diseases of lower animals.

Salmon and Smith, 1886, produced immunity in pigeons to the living hog-cholera bacillus by the injection of sterilized cultures of the germ. It has been shown in all diseases amenable to protective inoculation that immunity is produced by the dead as well as the living germ.

Kitasato produced immunity against the tetanus bacillus and its toxic products by the injection of the filtrate of cultures which contained only the soluble products of growth—toxins. He further demonstrated that the blood-serum of the immunified animal protected the susceptible.

These experiments, with those of Behring and Roux on the production of immunity to the toxins of the bacillus of diphtheria, were the beginning of antitoxins and serum-therapy.

The greatest value of serum-therapy is its prophylactic property. To protect the body after a pathological process has begun is difficult, but it has been accomplished. The obstacle in attaining success as a curative agent is the fact that it takes two to three days to acquire immunity. The principles of serum-therapy have been most thoroughly investigated in diphtheria and tetanus by Behring and Kitasato. The subject will receive further consideration under the special bacteria.

PART SECOND.

BACTERIOLOGICAL TECHNIQUE.

(a) General Principles of Sterilization and Disinfection. (b) Preparation of Culture Media. (c) Methods of Isolation and Culture. (d) Methods of Staining Bacteria. (e) Methods of Observing and Differentiating Bacteria.

Part Second.

BACTERIOLOGICAL TECHNIQUE.

(a) GENERAL PRINCIPLES OF STERILIZATION AND DISINFECTION.

It is of paramount importance in bacteriological research that all apparatus and culture media shall be perfectly free from germ life. The process of getting them into this state is spoken of as *sterilization*.

The application of *moist heat, dry heat, filtration,* and *chemical agents* are the means employed to accomplish this end. It is obvious that chemical agents are not applicable to the sterilization of culture media for the reason that the addition of these agents would render the media unsuitable to bacterial development.

Heat is applied either in the *dry* or *moist* state. Dry heat may be used for test-tubes, glass dishes, flasks, etc., intended to receive culture media, or, in other words, in the sterilization of all objects upon which it would not have a detrimental influence. Objects that can withstand direct heat (platinum needles, knives, etc.) are sterilized by passing them through the flame of a Bunsen burner or alcohol lamp.

In **Dry Sterilization** the objects, previously washed and dried, are placed in the oven and heated for an hour at 150° C. These hot-air ovens are made of sheet iron, and have double walls with tubulatures for thermometers and thermo-regulators. Loose cotton plugs are inserted into the test-tubes and flasks before sterilization. They act as air filters and

(33)

prevent the entrance of germs after sterilization. Browning of these plugs is taken as an indication that all bacteria and spores within the cotton and tubes have been destroyed.

Dry heat is not applicable in the sterilization of fluids, as these would rapidly evaporate in the hot-air oven. This method is not used so much as formerly.

Moist heat is used principally for sterilization of culture media. For the application of this method, which consists in exposure to steam at 100° C., several different forms of apparatus are in use, preference being given to the steam chest devised by Koch and the Arnold and Boeckman steam sterilizers.

Koch's apparatus is a large copper cylinder covered by felt, and divided by a diaphragm into two compartments. The lower chamber is filled with water, which is heated from below by a large burner. The objects to be sterilized are placed upon the diaphragm, where they are exposed to streaming steam. An aperture in the lid allows the escape of excess of steam.

The Arnold steam sterilizer is equally as effective as Koch's and much cheaper. It is made of tin or copper, and consists of a covered sterilizing chamber which rests over an open pan filled with water. In the bottom of this pan are perforations which allow the water to slowly drain into a lower chamber. This lower compartment is next to the source of heat, and has a tube extending from it through the water pan to the sterilizing chamber above. Through this tube the steam is conducted to the sterilizing chamber, where it condenses and drops again into the water pan below.

The temperature of the sterilizing chamber reaches 100° C. This is sufficient to destroy in fifteen minutes all fully developed bacteria, but not spores. To insure perfect sterility the culture medium is now allowed to remain for twenty-four hours at room temperature. Any spores which escaped destruction, in this time grow out into vegetative cells. The steaming is repeated and these are destroyed. For additional safety the procedure is carried out again on the third day. The medium is now sterile, and will remain so indefinitely, provided no bacteria gain access from without. This method of destroying germ life is spoken of as *discontinuous* or *intermittent sterilization.*

For rapid sterilization steam under pressure is used. In the various autoclaves a single exposure to 115° C. for twenty minutes will, as a rule, be sufficient to destroy germ life. This method is used only in the preparation of bouillon and agar-agar; gelatin, so prepared, loses entirely its solidifying property.

Sterilization by filtration is useful for the separation of bacteria from the soluble products of their growth, *i. e.,* toxins. The method is of no value in the preparation of culture media. Various arrangements of the Pasteur-Chamberland filter are used.

Disinfection is the destruction of disease-producing germs by heat or chemical agents. It is consequently not so wide a term as sterilization, which implies the destruction of all forms of bacterial life. The agents used are spoken of as *disinfectants* or *germicides.** They act by forming compounds with the protoplasm of the bacterial cell. To be effective it is therefore necessary that they shall come in actual contact with the bacteria. It follows also that they should be in solution. Spores are more resistant to the action of germicidal agents than fully grown cells. Germicides are more effective at high temperatures (35° to 40° C.).

Antiseptics are agents which restrain or prevent the growth of bacteria without destroying them. The germicidal power of chemicals differs for different species and with the same species under different conditions. A substance may be

---

* In the process of disinfection it oftentimes becomes necessary to destroy entirely the infected material. Sterilization, on the contrary, produces no change in the character or composition of material submitted to it, but merely destroys the bacteria.

germicidal for one species and antiseptic for another. Occasionally one meets with a culture of some micro-organism which shows much greater power of resistance than others of the same species. Many disinfecting agents which destroy pathogenic bacteria have less effect upon saprophytes. Bacteria can be made tolerant of the presence of a germicide by gradually accustoming them to its presence.

While many substances known to us are able, in extremely small amounts, to prevent the growth of or to destroy bacteria, their use for this purpose is not possible in the treatment of disease for the reason that inert compounds are formed as readily with the animal tissues as with the protoplasm of the bacterial cell. Thus, bichloride of mercury introduced into a part to destroy the bacteria forms with the tissues a harmless albuminate upon which the bacteria may feed. Germicidal agents are all poisons to the animal economy, and if given in sufficient quantity to destroy the bacteria in the diseased organ would cause death of the individual.

Methods of Determining Antiseptic and Germicidal Values.—The points to be determined are, *first*, the amount of the agent in proportion to a known quantity of culture fluid which is required to prevent the development of bacteria without destroying them ; *second*, the amount of the agent in solution necessary to destroy them within a given length of time — usually two or four hours.

Two methods may be used. In the experiments of Sternberg " a certain quantity of a recent culture (usually five cubic centimeters) was mixed with an equal quantity of a standard solution of the germicidal agent," and after a certain length of time a platinum loop full of the mixed culture and germicide was transferred to a fresh sterile tube. If the germicidal solution was in the proportion of 1–100, and if this, added to an equal quantity of a liquid culture, was the smallest amount required to restrain development, the *antiseptic power* of the agent was set down as 1–200.

Tests for the *germicidal power* are made in the same way; the determination being made of the minimum strength of the germicidal agent which, when mixed with equal parts of a liquid culture of the test organism, was sufficient to destroy it in a given length of time.

Another method of determining antiseptic and germicidal values consists in introducing the test organism into solutions of the germicidal agent in known strengths, and by making cultures from different strengths of the germicidal solution. The least proportion in which it will restrain development of or destroy the test organism within a certain length of time represents its antiseptic and germicidal powers.

(b) PREPARATION OF CULTURE MEDIA.

The culture media most in use are bouillon, nutrient gelatin, nutrient agar-agar, and blood-serum.

Bouillon.—Prior to the introduction of solid media, bouillon was used almost in exclusion of all other substances. The mode of preparation is as follows:

To 500 grams of fresh, finely chopped lean beef is added a liter of water. It is allowed to stand in a cool place over night, and strained through cloth until a liter of the infusion has run through. If there be less than the desired quantity, the deficiency can be supplied with water. To this liter of beef infusion 5 grams of chloride of sodium and 10 grams of dried peptone are added, and the mixture boiled for a short time in order to coagulate the albumin.

The acidity, due to the presence of sarcolactic acid, is neutralized by the addition of a saturated solution of sodium carbonate or potassium hydrate, drop by drop, until the reaction, tested by litmus paper, is slightly alkaline. The bouillon must be boiled again for half an hour to coagulate any alkaline albumins which may have been formed during the process of neutralization. The next step is to filter a small quantity into a test-tube; let cool, then boil again, to see if it remains clear. If not, it must be boiled and filtered as before. If this precaution is not taken, there is likely to result a turbid fluid in taking the next step in its preparation.

When the bouillon remains clear after this test it is filtered directly into sterile test-tubes to the depth of about an inch. Larger quantities are kept in conical flasks—Erlenmeyer's.

We are now ready to complete the preparation by sterilizing it. This is necessary, because during filtration contamination was unavoidable. Thorough sterilization is obtained by the fractional method above described, remembering to have the cotton plugs in place.

In many laboratories bouillon is made by using 2 grams of Liebig's or Armour's Beef Extract instead of fresh meat. The bouillon thus made serves as well as the other, and has the advantage of not requiring to stand twelve or twenty-four hours. The other steps are the same as in making bouillon from meat.

A cloudy precipitate, consisting of phosphates, may form after the solution cools, but can be removed by letting the bouillon cool and filtering before sterilization. The medium thus prepared is a clear, straw-colored fluid.

Nutrient Gelatin is prepared by adding to a liter of bouillon 100 grams of French gelatin, heating to dissolve the gelatin, neutralizing, boiling for about twenty minutes and filtering. It is essential that the filter paper be accurately plaited, and that it be moistened with hot water before pouring on the gelatin. If the filtrate comes through clear and does not form a precipitate upon heating, it is put into test-tubes and flasks. It should not be allowed to come in contact with the wall of the test-tube near the mouth.

Should heating cause the formation of a precipitate, the reaction should again be tested and if necessary corrected. The precipitate is removed by adding the white of an egg, boiling for a short while and filtering. Before the medium is ready for use it must be sterilized by the fractional method.

When cold it forms a transparent, straw-colored, gelatinous mass which is solid at all temperatures below 25° C.

Care should be exercised in the preparation of gelatin that it is not exposed to prolonged high temperature, as this destroys its property to gelatinize.

For the summer months fifteen instead of ten per cent of gelatin may be used.

Nutrient Agar-agar.—In the preparation of this medium agar-agar* is substituted for gelatin. This substance, which comes to us in dried strips, is added to neutral bouillon in the proportion of 20 grams to a liter. The mixture is kept in a water bath until the agar is dissolved (about six hours) and is then neutralized, boiled for half an hour, and passed through properly folded, moistened filter paper.

As the filtration of agar through paper is a tedious procedure, it has been suggested to filter through absorbent cotton. The cotton, placed in the bottom of a funnel, is moistened with hot water and the medium poured upon it, the filtrate being received in a flask. The funnel and flask are set in a steam sterilizer at low heat, and the medium is allowed its time to filter. The method is more rapid and as satisfactory as filtration through paper.

Test-tubes are filled with the medium to the height of about an inch. The tubes and the agar remaining in the flask are sterilized by the fractional method. Prolonged heating has no effect upon the gelatinizing property of agar.

When agar has congealed it is firmer than gelatin, somewhat darker in color, and not quite so transparent. It remains solid below 45° C., and is not liquefied by any known bacterium.

Glycerin Agar-agar. — As certain micro-organisms (tubercle bacillus, diphtheria bacillus) prosper better in the presence of glycerin, most laboratories have tubes of agar on hand to which from three to eight per cent of glycerin has been added. In appearance it differs little from ordinary agar.

Agar-Gelatin.—The medium known by this name is 5 per cent nutrient gelatin, to which 0.75 per cent of agar-agar has been added. It has a higher melting point than gelatin.

---

* Agar-agar is the native Cingalese term for several species of algæ found upon the shores of Ceylon. It is used by the Brahmins and Chinese for much the same purposes as gelatin is with us.

Blood-Serum is obtained by catching, in a sterile deep vessel, fresh blood from an animal, allowing it to stand under cover for twenty-four hours, and drawing off the serum with a pipette. It is put into tubes which are heated (in an oblique position), in a special apparatus devised by Koch, to 68° to 70° C. until firmly coagulated. It is subsequently sterilized by the fractional method at low temperature.

The medium prepared in this way has a yellowish-white color, is translucent, and has a firm consistence. If higher temperature be applied in coagulating the medium it becomes more compact and opaque.

Blood-serum is sometimes prepared in a similar manner from placental blood. The fluid of hydrocele and ascitic and pleuritic exudations have also been used.

It is sometimes desirable to preserve blood-serum in a liquid state as a culture medium and for other purposes. It may be sterilized by heating to 50° C. for one hour on five or six consecutive days. Another method of preservation recommended consists in the addition of a small quantity of chloroform and keeping the serum in an air-tight flask. When it is to be used the chloroform can be expelled by heating to 50° C. on a water bath.

Loeffler's Blood-Serum Mixture consists of one part of bouillon containing one per cent of grape sugar, and three parts of blood-serum prepared according to the method just given. It is used almost exclusively for the cultivation of the diphtheria bacillus.

NOTE.—Any of the solid media mentioned can be bought in tubes ready for use.

Potato Cultures.— A potato which has been thoroughly scrubbed and washed in a solution of corrosive sublimate 1-500, is cut in halves with a sterile knife. The pieces are placed, cut surfaces upward, in sterile dishes with covers.

Or by means of an apple-corer a cylinder may be taken from a potato and divided obliquely, each part being put into a test-tube with cut surface upward. The tubes or dishes should subsequently be sterilized by the fractional method. Where an acid medium is required, young potatoes, having a more acid reaction, should be preferred.

Holz's Potato Gelatin is a combination of gelatin and potato juice. The juice is obtained from potatoes by grating and allowing them to set over night and straining. This is filtered after twenty-four hours. To 400 grams of the filtrate are added 40 grams of gelatin. It is boiled slowly three quarters of an hour, filtered, and drawn into proper vessels, then sterilized by fractional sterilization. The medium is clear, transparent, and has a brownish color.

Elsner's Medium is Holz's potato gelatin modified by the addition of 1 per cent of potassium iodide. Elsner advises that the medium should have such a degree of acidity that 10 c.cm. are neutralized by 2.5 to 3.0 c.cm. of a decinormal solution of caustic soda or potash.

Besides the media mentioned there are a number of others of minor importance, such as milk, rice-milk, urine, bread-pap, etc.

Use of the Different Media. — Bouillon is especially adapted to the keeping of pure cultures.

The property of gelatin to solidify at room temperature gives it decided advantages over bouillon in the isolation of bacteria. It is also used to differentiate those bacteria which have the property of liquefying gelatin from non-liquefying species. Some organisms form a characteristic growth upon it. Its transparency fits it admirably to the study of the growth of pure cultures. For the cultivation of parasitic species it has no advantages over bouillon, as it liquefies at from 24° to 26° C.

For the forms which require a higher temperature agar is the preferable medium.

Blood-serum has the advantage that it furnishes nutrient elements in about the same proportions that they are met with in the animal economy. It is used for the cultivation of such species as require concentrated nourishment.

The disadvantage of blood-serum is its opacity, and that when once solid it can not be again liquefied. This makes it unfit for plate cultures according to the method of Koch.

Potatoes were formerly used extensively, but have been largely replaced by the other solid media. They are valuable as a diagnostic means, certain micro-organisms growing upon potatoes in a characteristic manner. Chromogenic bacteria grow and produce color most luxuriantly upon them.

For the purpose of differentiating bacteria the addition of various chemicals to culture media is useful. We find certain forms that prosper best in the presence of certain chemicals, *e. g.*, glucose, glycerin, chloride of sodium, etc. Others produce reactions during their growth with the chemicals by which the species can be recognized. For this purpose litmus (in sterile milk) and some of the aniline colors are used to detect the production of acids or alkalies. In the same way nitrite of potash is used to test the nitrifying property, glucose the fermenting property of the growth, and rosalic acid the production of acids.

(c) THE METHODS OF ISOLATION AND CULTURE.

Bacteria growing in a liquid medium must necessarily intermingle, consequently a separation of the different species by transplantation from one liquid medium to another is practically impossible. The ingenious idea of solidifying the medium by the addition of gelatin, thereby introducing an easy means of isolation, originated with Robert Koch.

The method of isolation is based upon the principle of dilution and distribution of the germs in the medium while it is in the liquid state, and allowing subsequent solidification upon a flat surface so that each germ will have a field of its own in which to grow.

Koch's Plate Method, as originally introduced, is seldom carried out at present, though the principle of all the modified methods is the same. The procedure suggested by Koch is as follows: The gelatin or agar of three tubes is liquefied by gentle heat. Then with a sterile platinum loop or needle (made by inserting about three inches of platinum wire into a glass rod) a minute portion of the material containing the mixture of micro-organisms is passed into the first tube. As the quantity of material introduced, if ever so small, would not be sufficiently diluted, the needle is, after stirring it about thoroughly in the medium of the first tube, carried into that of the second. This procedure is repeated three times. The needle is then carried from the second to the third tube two or three times. During these manipulations the tubes, previously numbered, are held between the fingers of the left hand, as are also the cotton plugs. That portion of the plug which fits in the tube should be carefully guarded against contact with external objects. The germs in the third tube are, after the procedure mentioned, in a very diluted state.

The contents of the tubes are next poured upon three sterile glass plates numbered to correspond with the tubes. The medium is spread with the lips of the test-tubes previously sterilized in a flame. The plates are kept on a tray resting on cracked ice, and when the medium has congealed are put away in a covered glass dish for colonies to develop.

Petri's Modification.—In this method shallow glass dishes (diameter 120–150 mm.) with tops are used. They have almost replaced Koch's plates. The dilution is carried out precisely as in the foregoing method. The tubes are then emptied into sterilized dishes. The method is much simpler than Koch's, and meets all requirements. Besides taking up less room in the incubator, the Petri dishes have the advantage over Koch's plates in that the colonies can be observed (by turning the dish bottom upward) without removing the cover.

To apply the method to blood-serum the medium may be solidified in the dishes during its preparation. The material supposed to contain bacteria is smeared over its surface, and as the colonies develop they are transferred to tubes. Blood-serum plates are of value in the isolation of the gonococcus.

Von Esmarch's Tubes.—To avoid the danger of contamination during transfer of the medium from tubes to plates, and to simplify the method of isolation, Von Esmarch introduced a means by which the walls of the original tube are converted, so to say, into plates. The tubes should contain about one half the usual quantity of medium.

After inoculation and dilution in the usual manner the cotton is replaced in the tubes, and they are rolled on a piece of ice or under the hydrant until the gelatin congeals in a thin layer along the walls.

The disadvantage of the test-tube method is that liquefying forms cause the gelatin to run together. It can not be carried out so well with agar, as this medium usually gravitates to the bottom of the tube. By placing the tubes in a slanting position this difficulty can be obviated to a certain extent.

Appearance of Colonies.—After twenty-four hours, more or less, the surface of the medium in the Esmarch tubes or on the plates will appear specked with minute masses. These are colonies or families of bacteria, each representing a single species — unless, by chance, two or more bacteria were thrown in close apposition to each other.

Under the microscope with a low power (60 diameters) the colonies vary in appearance, some being round, others irregular, some having rough or serrated, others smooth edges, etc. Some grow above the surface of the gelatin, forming elevated patches; others liquefy the medium and sink into the depressions. They may be variously pigmented, or colorless and barely visible to the naked eye. As the colonies grow larger they come in contact with each other, but show no tendency to coalesce.

Fig. 7.

For their further development it is essential to transfer them to a larger field of nutriment as soon as possible. This is done under a low power of the microscope. The point of a sterile platinum needle is plowed through a colony and the material procured transferred to a new tube. It is allowed to grow in the tube— tube culture; or a new plate may be poured—plate culture. The growth will represent a single species of micro-organism, and is spoken of as a *pure culture.*

Test-tube Cultures are made either on the surface of the medium—stroke or smear culture; or in its depth— puncture or stab culture.

A *stroke culture* is made with a single stroke of the needle, from below upward, on the surface of the medium, which for this purpose must have an oblique surface. **Agar** is most frequently employed for surface cultures, gelatin being of use only for the cultivation of non-liquefying forms which grow at room temperature. Stroke cultures vary as widely as the colonies on plates in rapidity of growth, shape, color, etc.

A *stab or puncture culture* is made by passing the needle deep into the medium, the surface of which should be at right angles to the sides of the tube.

The growth takes place along the line of puncture. It shows characteristics in certain species not recognizable in the surface culture. Filamentous outshoots may extend in a radiating manner from the line of puncture. Isolated masses may be distributed throughout the gelatin near the track of the needle. The gelatin may be liquefied in a characteristic way (Fig. 7), and pigments may be deposited which distinguish certain forms, etc.

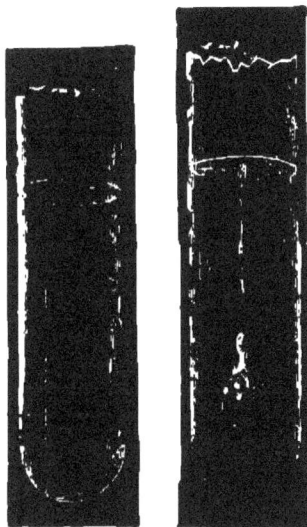

Fig. 8.

The method also differentiates between aerobic and anaerobic varieties. If the growth occurs only near the surface, we have to deal with an aerobic variety. If it takes place only in the depth, it is an anaerobic variety. (See Fig. 8.) Growth both on the surface and along the needle track denotes a facultative anaerobe.

Cultivation of Anaerobic Bacteria.—For that variety of micro-organisms which prosper only when all oxygen is excluded, various methods of cultivation have been suggested.

A very simple and practical means is that of Liborius. He heated a tube almost full of agar, containing 2 per cent dextrose, to expel as much oxygen as possible, and after allow-

ing it to cool to about 40° C. inoculated the material to be
examined into the depth of the tube and allowed the medium
to congeal. Colonies of anaerobic organisms grew near the
bottom of the tube, the aerobic forms developed near the
surface. The colonies were examined by breaking the
tube and making sections of the agar for examination or
further culture.

Eggs for Anaerobic Culture (Hueppe). — Fresh eggs have
been suggested for the cultivation of anaerobic forms. After
they have been thoroughly cleansed in a sublimate solution,
rinsed and dried, they are inoculated by passing a needle
through a small hole in the shell. The opening is closed
with paper and collodion.

For the development of anaerobic forms as surface cultures
or in liquid media their surroundings must be freed of
oxygen. A number of methods have been suggested to
accomplish this, some based upon the principle of exhausting the air by mechanical means, others of replacing it with
inert gases (H and CO_2), and others of absorbing with chemicals the oxygen of the air in which the cultures are grown.

The method of exhausting the air, which was accomplished by
means of an air-pump, has not proven very satisfactory, as
a quantity of oxygen always remains in the medium, and
is detrimental to the development of anaerobic bacteria.

The method of absorption of the oxygen (Buchner) is simple
and reliable, and answers the purpose admirably for test-
tube cultures. The tube containing the material for development is plugged with cotton and set into a larger tube
containing a mixture of 1 gram of pyrogallic acid and
10 c.cm. of a 10 per cent solution of caustic potash, and the
latter tightly closed with a rubber stopper. The small tube is
kept above the deoxidizing material by means of a wire
bracket. (Fig. 9.) The pyrogallic acid takes the oxygen
from the air and furnishes the condition necessary for the
development of anaerobic germs.

A method combining the principle of absorption with that of displacing the air with hydrogen is an excellent way of growing anaerobic surface cultures. It is the preferable method for plate cultures. The plates are placed under a bell-jar which is filled with hydrogen gas. The bell-jar is set into a dish containing liquid paraffine and the hydrogen passed into it through a tube and allowed to escape through a second tube until it comes away pure, when both tubes are removed. To absorb any oxygen which might remain, a shallow dish containing the alkaline pyrogallic acid solution of the previous method is placed under the jar.

Fig. 9.

These methods are applicable to the cultivation of anaerobic species in both liquid and solid media.

Incubators and Thermo - regulators. — The proper temperature at which to keep cultures for their growth differs with the species. Some prosper better with low heat, and are cultivated at room temperature (20° to 25° C.). Others (particularly the parasitic species) require a temperature near that of the body (35° to 40° C.), and are cultivated in an incubating oven.

An incubator or thermostat, as these ovens are commonly referred to, is a double-walled box made of copper, between the walls of which is a space for water. Externally it is covered with some non-conducting material, usually felt or asbestos. The temperature in the oven is dependent on the temperature of the water between its walls. The water is heated by a burner from below, the gas supply being governed by an automatic thermo-regulator.

(d) METHODS OF STAINING BACTERIA.

Material aid has been given bacteriological research by the introduction by Weigert (1876) of the basic aniline dyes as staining agents. These stains, which are coal-tar derivatives, are on the market in various shades and qualities. Of those most in use may be mentioned methylene blue, methyl violet, gentian violet, fuchsin, Bismarck brown, and vesuvin.

Preparation of Staining Solutions.—Saturated alcoholic solutions of the dyes are known as *stock solutions.* Such solutions of the dyes most in use should be kept on hand. They are not used for staining, but various dilutions and compound solutions are made from them. Dilutions are made with water for staining bacteria on the cover-glass.

Double Staining.—A combination is sometimes made of the basic dyes, which stain the nuclei of tissue cells and bacteria, with diffuse stains, such as eosin and picric acid, which stain connective tissue, the protoplasm of the cell body, etc., thus giving a contrast of colors.

An example of such combination is found in the following solution :

CZENZYNSKI'S SOLUTION.

Sat. aqueous solution methylene blue ....20 c.cm.
One half per cent sol. eosin in 70 per cent alcohol.10 c.cm.
Water ....40 c.cm.

This is a permanent solution.

Mordant Solutions.— Agents used to fix or increase the intensity of stains are called mordants (*Beizen*). Some act by giving to the stain greater power of impregnation. Examples of such are aniline oil, carbolic acid, and caustic potash. Others act by forming compounds with the stains whereby the color is fixed. Of these, iodine and iodide of potash in

Gram's method, and tannic acid and ferrous sulphate in Loeffler's method for flagella, are examples.

The following are some of the principal mordant solutions:

GRAM'S SOLUTION.

Iodine.................................... 1 gram.
Iodide of potash .... 2 grams.
Water ....300 grams.

This solution deteriorates with age.

LOEFFLER'S SOLUTION FOR FLAGELLA.

20 per cent sol. tannic acid in water ....10 c.cm.
Aqueous sat. (cold) sol. ferrous sulphate....... 5 c.cm.
Alcoholic sat. sol. fuchsin, or methylene blue .. 1 c.cm.

This solution improves with age.

LOEFFLER'S MORDANT SOLUTION (BUENGE'S MODIFICATION).

Saturated watery sol. of tannic acid ....30 c.cm.
Watery sol. sesquichloride of iron, 1–20.......10 c.cm.

To ten parts of this mixture is added one part of saturated watery solution of fuchsin. This solution improves with age. It should not be used for eight days after preparation.

EHRLICH'S ANILINE WATER SOLUTION.

Aniline oil .... 4 c.cm.
Distilled water............................100 c.cm.
Shake thoroughly and filter through moist filter paper
      (Aniline water).
Add stock solution of an aniline dye ....10 c.cm.

This solution spoils in four to six weeks.

ZIEHL'S SOLUTION.

Powdered fuchsin .... 1 gram.
Alcohol................................. 10 grams.
Carbolic acid........................... 5 grams.
Water.................................100 grams.

This solution is permanent.

LOEFFLER'S METHYLENE BLUE SOLUTION.

Stock solution methylene blue .... 30 c.cm.
1–10,000 watery solution caustic potash.......100 c.cm.

This is a permanent and useful stain.

KUEHNE'S METHYLENE BLUE SOLUTION.

Methylene blue .... 1½ grams.
Absolute alcohol ....10 c.cm.

Triturate and add slowly 100 c.cm. of a 5 per cent watery solution of carbolic acid.

The application of gentle heat to the solutions, while acting on the specimen, also enhances their impregnating property.

Decolorizing Solutions.—A number of agents may be used to remove surplus stain, notably the mineral acids, acetic acid, alcohol, and water.   Guenther combines two of these bleaching agents:

GUENTHER'S DECOLORIZING FLUID.

Hydrochloric acid .... 3 c.cm.
Alcohol ....100 c.cm.

The bleaching agent may be combined with a staining fluid whose object is to give a contrast color to the bleached tissue.   An instance of this is

GABBET'S ACID BLUE SOLUTION.

Methylene blue ....2–3 grams.
20 per cent sulphuric acid ....100 c.cm.

Staining on the Cover-Glass and Slide. — Cover-glasses and slides should be made of clear white glass.   They should be rendered absolutely clean by dipping them into a weak acid solution, washing in alcohol, and drying with a soft cloth.   After they have been cleansed the fingers should only come in contact with the edges.

Liquid material (pus, blood, etc.) and pure cultures are stained on the cover-glass or slide.

Pus, blood, and such material is spread in a thin layer on the glass and dried in the air.

Pure cultures and mixtures of bacteria in fluids are prepared by transferring to the glass with the platinum loop a small drop of the fluid containing them, and allowing it to dry in the air.

When the micro-organisms are densely crowded, as they are in pure cultures on solid media, and sometimes in liquids, a very minute portion is diluted on the glass with a drop of water and allowed to dry.

Fixing.—After the specimen has thoroughly dried fixation may be accomplished either by placing it for twenty-four hours in a mixture of equal parts of ether and alcohol or by applying dry heat, which coagulate its albumin.

The Method of Fixing by Heat.—Grasp the edges of the cover-glass or slide between thumb and fingers, and pass it, smeared or "buttered" side up, three times through a low flame of a Bunsen burner or an alcohol lamp. It would be difficult to describe the rapidity with which this should be done. It must be acquired by experience. Insufficient heat would cause incomplete fixation, and the material would be lost in the staining process. Too much heat would dry up the cells and rob them of the property of taking the stain.

In order to avoid such errors, it has been suggested to hold the glass between the fingers, which act as a thermometer during fixation. The heat that the fingers are able to stand will do no harm. After fixing, the specimen may be stained at once or it may be kept indefinitely.

Simple Methods of Staining.—For staining slide and cover-glass specimens simple stains are usually employed, preferably methylene blue or fuchsin. The stock solutions are diluted with water (1–100), and a drop or two of this is brought upon the dried and fixed film. After one to three minutes it is washed in clear water and is ready to be examined. Owing to the variation in the strength of dyes and the difference in the affinity of species for stains it is not wise to adhere to any fixed rule as to the length of time the stain shall be allowed to act. The intensity of action must be studied and dilution made accordingly. The specimen may now be mounted in water or in Canada balsam.

If it is desired to study a specimen prepared on the cover-glass, a drop of water should be placed in the center of the slide and the cover-glass placed upon it, stained side down. If the film has been stained upon the slide, a drop of water is placed upon it and the cover-glass applied. If, after examination in water, it is desired to mount in balsam for preservation, the water is thoroughly evaporated, a drop of balsam placed on the slide, and the cover-glass lowered upon this.

These specimens were formerly prepared entirely on the cover-glass. As most preparations are made merely for diagnostic purposes, staining on the slide will be found much more convenient, and is now largely practiced. It does away with forceps, is cleaner, and on account of less breakage is more economical.

Staining Bacteria in Blood.—A microscopical examination of the blood of living individuals is sometimes desirable for the detection of the bacteria which cause septicemia, as well as for the spirillum of recurrent fever and other organisms which sometimes find their way into the circulation. In these examinations there are certain procedures which require special attention.

The blood is secured from a sterilized finger-tip or lobule of the ear by the prick of a needle. A drop is brought upon a clean cover-glass. To spread the material evenly another glass is dropped upon the first. They are carefully drawn apart and allowed to dry in the air.

Special methods of fixation are required in order to preserve the red blood-corpuscles. Ehrlich recommends heating the dried film to 120° C. for one hour in the hot-air box. Exposure to formaldehyde vapor is also used, or the specimen may be floated, spread side down, for ten minutes upon a mixture of equal parts of absolute alcohol and ether.

Specimens prepared in this way can be stained with the simple aniline dyes. Methylene blue should be given the preference, as it stains lighter and brings out the bacteria well. In deeply stained specimens blood-corpuscles and plasma may hide the bacteria.

Guenther has suggested a method by which the bacteria alone are stained. After fixing he washes in a one to five per cent solution of acetic acid. This takes the hemoglobin out of the blood-corpuscles and dissolves most of the plasma adherent to the glass. The bacteria remain unchanged, and can, after drying, be stained in the usual manner. If the plasma is very dry and fails to come off, Guenther treats it with two per cent solution of pepsin. This dissolves the plasma but not the micro-organisms.

For purposes of demonstration, Czenzynski's method (see staining methods) is preferable, as the double stain makes neater specimens.

Smear preparations may be made on slides or cover-glasses from the blood and organs after death, observing the same technique.

Adhesion or Contact Preparations.—For study of the arrangement of bacteria in colonies a cover slip is dropped lightly upon an isolated plate colony. When it is raised the growth adheres to it. It is dried in the air and stained. This is a contact preparation called by the Germans "Klatsch-praeparat."

Staining of Spores.*—For the staining of spores several methods have been devised. The exosporium is ordinarily impenetrable to dyes and requires special treatment. It has been shown that spores can be stained by previously exposing

---

* In those methods which require prolonged action of the stain the preparation (cover-glass) is floated, spread side down, upon the staining fluid in a watch glass.

them to high temperature, which can be accomplished by passing the cover-glass through the flame ten or twelve times. An hour in the hot-air oven at 120° C. will also answer the purpose. After this treatment they take the ordinary stains, while the bacteria themselves do not absorb the dyes so well. A very satisfactory method for staining spores is that of Neisser, which is as follows:

1. Float the cover-glass in Ziehl's solution or Ehrlich's aniline water solution.
2. Heat until steam arises, then set aside for half hour.
3. Bleach in 3 per cent solution hydrochloric acid in alcohol.
4. Wash in water.
5. Counterstain in weak watery methylene blue.

Spores retain the stain after the cell body has been decolorized. The latter stains blue.

The method of Moeller is as follows:

1. Dry and fix by passing through flame.
2. Place in chloroform...................... 2 min.
3. Wash in water.
4. Place in 5 per cent chromic acid...........½–2 min.
5. Wash in water.
6. Stain as in Neisser's method.

For staining of spores fresh spore-bearing rods are most suitable.

Staining of Flagella.—The demonstration of flagella is extremely difficult. It is important that the cover-glass should be previously freed from oil and other foreign matter by washing in alcohol. The minutest quantity of a fresh culture is stirred into a drop of sterile distilled water on a cover-glass. The water is allowed to evaporate and the film fixed, care being exercised not to apply too much heat. Loeffler devised a method which is now generally employed:

1. Heat gently in Loeffler's mordant solution; move the specimen about in the solution for one to two minutes.
2. Wash in water.
3. Wash in alcohol until all trace of the mordant is removed.
4. Stain in neutral Ehrlich's solution, heated until it steams, and leave set for one minute.
5. Wash in water.
6. Dry.
7. Mount.

The method as it is given will, according to its discoverer, stain the flagella of but few varieties of bacteria.

For staining some species an alkali, for others an acid must be added to the mordant. Which to use, and how much, must be judged empirically. This makes the method difficult of execution and unsatisfactory.

Buenge's Modification of Loeffler's Method is carried out in the same way as Loeffler's, with the exception that the author uses a different mordant. (See staining solutions.) It has the advantage that no alkali or acid need be added to the mordant, and is said to be more satisfactory than Loeffler's method.

Special Methods of Staining Bacteria on Cover-Glass and Slide. — The iodides used in Gram's method form a compound with the aniline dyes and the mykoprotein of the cells upon which alcohol has a very slow solvent action. By virtue of this action the bacteria are, during the process of bleaching the previously overcharged specimen, the last to give up the stain. As the process of bleaching with alcohol alone was unsatisfactory, Gram's original method was modified by Guenther, who added an acid to the alcohol in bleaching.

Gram's Method is as follows:

1. Stain in Ehrlich's solution ....1– 5 min.
2. Drain off superfluous stain with filter paper.
3. Gram's solution........................1– 2 min.
4. Guenther's acid alcohol.................8–12 sec.
*5. Bleach in alcohol ....2– 6 min.
6. Wash in water.
7. Dry in air.
8. Mount in balsam.

Specimens treated according to this method show unstained tissue elements having a yellowish hue, in which the bacteria are seen as dark blue or red bodies, accordingly as blue or red stains were used. The tissue cells can after the fifth step be contrast-stained with weak solutions of eosin or Bismarck brown.

Gram's method for the differentiation of bacteria is of considerable importance. While a number of bacteria retain their stain, there are others which are bleached in alcohol as readily as the tissue cells. Examples of those which do not stain by this method are the gonococcus, bacillus of malignant edema, chicken cholera, typhoid bacillus, Friedlaender's pneumobacillus, influenza bacillus, glanders bacillus, bacillus of bubonic plague, etc.

Examples of those which retain the stain are the anthrax bacillus, tubercle bacillus, diphtheria bacillus, streptococcus, staphylococcus, pneumococcus of Fraenkel, etc.

Czenzynski's solution is useful when it is desired to show the relation of bacteria to tissue cells, and also for staining the plasmodium malariæ. The cover-glass is floated upon the stain ten minutes, washed in water, dried, and mounted in balsam.

Staining of Bacteria in Tissue.—To prepare tissue for staining it must, while fresh, be put into absolute alcohol,

---

* The length of time the specimen should remain in alcohol varies considerably, depending upon the thickness of the material, quality of stain, and whether or not the specimen is moved about in the alcohol.

which is changed several times during the first four or five days. After that, or if desired after a longer period, it is transferred to a mixture of equal parts of alcohol and ether for several hours, then into thin celloidin for twenty-four hours, then into thick celloidin. It is exposed to the air, and when of proper consistence is fastened to a cork with thin celloidin. It can now be cut or kept in 70 per cent alcohol until ready for use. Sections should be made as thin as possible.

Loeffler's Method, which is the most popular for staining bacteria in tissue, is carried out as follows:

1. Stain in Loeffler's methylene blue sol.......3– 5 min.
2. Bleach in 0.1 per cent solution acetic acid.10–40 sec.
3. Bleach in alcohol.
4. Dehydrate in absolute alcohol.
5. Clear in oil of origanum ....2 min.
6. Mount in balsam.

This is applicable to all species of bacteria. A better method, but one serviceable only for a limited number of species, is that of Gram.

Gram-Guenther Method is the same as for cover-glass preparations, with the exception that after bleaching in alcohol the sections are brought into absolute alcohol, oil of origanum, and mounted as in the previous method. The tissue will appear pinkish and the bacteria dark blue or red, accordingly as methylene blue or fuchsin was used in Ehrlich's solution.

The diffuse stain is the first given up in the process of bleaching (connective tissue, cell protoplasm, etc.), then the nuclear stain of the tissue cells, and lastly the bacteria. Knowing this, it is easy in examining a section to tell whether the bleaching process has been continued long enough or too long.

If a contrast stain is desired, after decolorizing in alcohol the section may be put for one minute into a weak watery solution of eosin, Bismarck brown, or safranin, after which it is transferred to absolute alcohol, cleared, and mounted in balsam.

Weigert's Method for section staining, which has found considerable favor, is a modification of Gram's:

1. Stain in Ehrlich's solution ....1-3 min.
2. Absorb surplus stain with filter paper.
3. Treat with Gram's solution... ....1-2 min.
4. Decolorize in a mixture of aniline oil, 2 parts,
    xylol, 1 part, until bleached ....2-8 min.
5. Clear in oil of origanum or xylol.
6. Mount in balsam.

Kuehne's Method, though rather complicated, is occasionally used for staining sections. The method is as follows:

1. Stain in Kuehne's solution.................½ hour.
2. Wash in water.
3. Decolorize in weak sol. hydrochloric acid,
    10 drops to 500 grams............a few seconds.
4. Immerse in a solution of lithium carbonate
    (8 drops sat. sol. carbonate of lithium to
    10 grams water)................·........a few seconds.
5. Wash in water.........................2-3 min.
6. Wash in alcohol slightly tinged with methy-
    lene blue ....1-2 min.
7. Dehydrate in aniline oil tinged with methylene blue.
8. Wash in pure aniline oil.
9. Wash in xylol to remove oil.
10. Clear in turpentine.
11. Mount in balsam.

Methods of staining applied to individual organisms, as tubercle bacillus, glanders bacillus, etc., will be given in considering these special bacteria.

(e) METHODS OF OBSERVING AND DIFFERENTIATING BACTERIA.

A few species of bacteria by virtue of characteristic shape or arrangement or their reaction to certain staining methods can be differentiated by direct microscopical examination of the material containing them. Usually, however, a systematic study of the biological and morphological characters is necessary to determine the variety under consideration. With these means of diagnosis we often combine a third, the study of the action of bacteria upon lower animals. With the different varieties in pure culture the characteristic properties of each can readily be investigated.

Of the three methods employed in differentiating species, most importance is attached to the study of the (1) **Biological or Physiological Characters.** Great variation exists in the appearance of colonies of the same species on the different media as regards their shape, consistence, rapidity of development, etc., and in other phenomena occurring during the growth. Some liquefy gelatin, others do not. Various pigments and gases — foul and aromatic — are given off, while some exhibit the phenomenon of phosphorescence.

(2) **The Morphology,** which includes not only the shape but the arrangement of the micro-organisms, is studied with the microscope. While no investigation is complete without a microscopical examination, very few are complete with it alone. It is essential that we have a good microscope. The instruments made by some of our American manufacturers are reliable, reasonable in price, and meet all requirements. Besides the dry objectives $\frac{2}{3}$ in. and $\frac{1}{8}$ in., the instrument should be equipped with an "oil immersion" system ($\frac{1}{12}$ in.). In the use of this system the surface of the cover-glass and the objective are connected by a drop of

cedar oil, which has the same refractive index as glass. By preventing the rays of light which carry the image from entering a different medium (air), it prevents diffraction and brings about clearer images. For the demonstration of stained bacteria in tissues, it is also essential that the microscope have a substage condensor and iris diaphragm to increase and regulate the volume of light.

We examine bacteria microscopically, either in their natural state or by staining them. In the living state they are studied by the method known as the *examination in the hanging drop.* Its execution, which is simple, is as follows: A small drop of distilled water is put upon a clean coverglass. Bacteria are stirred into it with a platinum needle. The glass is then inverted and placed upon a slide, hollow-ground upon one side for this purpose. The drop is suspended from the cover-glass in the concavity of the slide. To hold the glass *in situ* and prevent evaporation the edge of the circular depression in the slide is surrounded with vaseline before the cover-slip is placed upon it.

This examination must be made with *weak light, i. e.,* with the iris diaphragm partly closed. Before using the high power the edge of the drop is gotten into the center of the field with the aid of the low power. The high power lens (preferably dry system) is then brought almost in contact with the cover-glass, and slowly raised with the fine adjustment until the contents of the drop are brought to view.

The micro-organisms are seen in a state of more or less activity. This may be due principally to the vibratory tremor which is seen, under the microscope, in all insoluble particles suspended in water, which has been termed *Brownian movement.* It should not be confounded with the motility which is generated in the cell itself, but which is not inherent in all bacteria. In the hanging drop the voluntary motion is easily recognized by the bacteria darting across the field. They change their relative position to other objects and to each other. In the Brownian movements they remain in the same relative position, but spin or dance about very actively.

Besides these movements, a circulatory motion at times takes place in the fluid. All objects in the field move in one direction. To do away with the vibratory motion of the drop, the bacteria are sometimes brought upon the cover-slip in liquefied gelatin or agar, which is allowed to congeal.

While the hanging drop is especially adapted to the study of motility, it also gives us a means of observing the phenomena of sporulation and fission. To study these processes a platinum loop of a culture from a solid medium is stirred into liquefied agar or gelatin. A small quantity of this medium is made into a hanging-drop mount, which is sealed with paraffine. The bacteria being in a fixed position can be observed without difficulty. A special incubator has been devised into which the microscope containing the slide is set. It has a window in front for the light to enter, and an arrangement by which the slide can be shifted from without. The eye-piece projects through the roof of the chamber.

The hanging drop also shows the arrangement of the cells in their natural state.

For the study of the size and shape of bacteria the stained specimens answer better. These also give us, as we have seen, valuable information concerning the reaction of the bacteria to certain stains, differentiating in this way between some species. Stained preparations should be examined with the iris diaphragm open to admit plenty of light.

(3) **Study of the Action of Bacteria on Lower Animals.**—This valuable adjuvant to our means of differentiating bacteria should not be underestimated. There is such similarity in biological and morphological characters of certain species that their pathological action aids materially in differentiating between them.

The animals upon which these studies are principally made are the rabbit, guinea-pig, and pigeon, although the mouse, rat, dog, cat, pig, sheep, etc., are also useful.

Inoculation is usually practiced by introducing the germs under the skin, either with a hypodermic syringe or by slitting the skin and introducing the material with a platinum needle. At times injections are made directly into the blood-vessels, or they may be made into the pleural and peritoneal cavities or the anterior chamber of the eye. Feeding experiments are also frequently made for the purpose of testing the possibility of disease production by the alimentary canal. Likewise germs are dried and sprayed into the air and inhaled by the animal to test the possibility of infection through the lung. Any symptoms which develop should be carefully noted. Some produce characteristic alterations of temperature, nutrition, secretion, circulation, respiration, etc. Death takes place in a variable length of time, depending, if the disease be an acute one, upon the susceptibility of the animal, the virulence of the germ, the quantity, and the route by which it was introduced. The pathogenic action of bacteria is often widely different in different species of animals.

When we wish to test the disease-producing power of a bacterium, inoculations are made with pure bouillon cultures about forty-eight hours old. The quantity used for the rabbit and guinea-pig seldom exceeds one cubic centimeter subcutaneously. For smaller animals the amount is proportionately less.

Animal inoculation is often of very great value for diagnostic purposes. Those animals which have, by experiment, been shown to be most susceptible to a particular disease may be inoculated with material suspected to contain the specific germ thereof. So we find that by the inoculation of guinea-pigs we can induce tuberculosis, when the specific germ could not be discovered by microscopical examination or cultures from the same material.

We also practice inoculation to determine the virulence of a bacterium or of its toxic products. An example of the latter is the testing of diphtheria toxins upon guinea-pigs.

No matter for what purpose inoculation was practiced, the examination must be continued to discover the cause of death. A post-mortem examination is necessary for this purpose, and should be made as soon as possible. The implements used should be sterilized by holding them in the free flame. The parts from which cultures are to be made must be removed under all aseptic precautions, and pieces transferred by means of sterile platinum loops to culture media. Stroke or stab cultures can be made or plates poured. After the material for culture has been secured, cover-glass preparations can be made from the blood and various organs. Parts of organs intended for the study of bacteria in stained sections should be cut into small cubes and placed in absolute alcohol for twenty-four hours or longer. After this they are imbedded, cut, and examined, as previously described.

PART THIRD.

PATHOGENIC BACTERIA.

Part Third.

PATHOGENIC BACTERIA.

BACTERIA OF SUPPURATION.

Suppuration is a termination of inflammation dependent upon the presence of various micro-organisms. It has been demonstrated that suppuration follows the subcutaneous inoculation of sterilized cultures as well as the living germ. The pus-producing property resides in the bacteria themselves and not in their products, as was demonstrated by Buchner, who introduced the filtrate of sterilized cultures of a number of species with negative results. The most common pus-producing species are:

(a) STAPHYLOCOCCUS PYOGENES AUREUS.

Synonyms: Micrococcus of infectious osteomyelitis; golden staphylococcus.

The staphylococcus pyogenes aureus is a facultative saprophyte. It has been found in water and the soil, but is most abundant wherever men and animals congregate. Hence it is often met with in the air of public places, hospital wards, etc. It is found very commonly on the healthy mucous membrane of the mouth, vagina, urethra, intestinal tract, and upon the surface of the body. It is present in 80 per cent of carbuncles, abscesses, and other acute suppurations.

Morphology.—A small, non-motile coccus, about 0.87 μ in diameter, grouped in irregular masses; often occurring as a diplococcus or in chains of four or five members.

Growth on Culture Media.—*Upon gelatin plates*, at room temperature, the staphylococcus forms on the second day small, white, punctiform colonies. Under the microscope they appear round, granular, with well-defined edges, and have a light-brown color. On the third day surface colonies begin to take on the golden-yellow color and to liquefy the gelatin. After the second day they do not increase much in size.

Fig. 10.

Streptococcus and staphylococcus from pure culture.

In gelatin puncture cultures liquefaction takes place along the needle track and quickly extends to the walls of the tube. The cocci in the upper part of the liquefied gelatin sink to the bottom of the tube, appearing as a golden-yellow mass. The upper part of the medium becomes clear.

Upon agar plates the colonies are similar to those on gelatin; on the third or fourth day the characteristic color appears.

Upon oblique agar and blood-serum at 37° C. a thick, white layer is produced without liquefaction. Color production at high temperatures does not keep pace with the growth. It is best seen upon oblique agar at room temperature, although cultures are commonly seen which on the second day, in the incubator, have the characteristic color.

Upon potato a thick pellicle forms. The potato soon reacts acid and has a sour-paste odor.

Bouillon quickly becomes 'cloudy — twelve hours in the incubator.

Milk is coagulated with the production of lactic and butyric acids.

The staphylococcus will develop in media having a slightly acid reaction. It grows either with or without oxygen, and at temperatures above 18° C., best at 37° C.

Vitality.—Fresh gelatin cultures are killed by exposure to a temperature of 56° C. to 58° C. for ten minutes. When dry, however, a temperature of 80° C. is required for the same length of time.

This coccus is able to withstand drying ten days. Cultures in gelatin or agar, if kept moist, will retain their vitality for a year.

Method of Isolation.—A small quantity of pus is stirred into liquid gelatin or agar, which is poured into Petri dishes and allowed to congeal. In the usual length of time colonies will appear if the coccus is present, and can be transferred to tubes.

Staining.—It stains with the simple aniline colors, and can be demonstrated in tissue by Gram's method. Acid hematoxylin solutions are also applicable.

Pathogenesis.—*Man.*—Pure cultures rubbed into or injected under the skin cause circumscribed suppurative inflammation. The possibility of infection by this germ through the unbroken integument has been demonstrated, and it has further been shown that they penetrate through the hair follicles, sudoriparous and sebaceous glands. The coccus is found, either alone or associated with other micro-organisms, in about 80 per cent of carbuncles, abscesses, and wound infections.

Suppurations excited by it have a tendency to remain localized, and there is always more or less disintegration of tissue. Not uncommonly, however, the infection becomes generalized and abscesses form in different parts of the body, notably in the joints, endocardium, and voluntary muscles. This production of mutiple abscesses (pyemia) by the staphylococcus is more frequent than by other pyogenic micro-organisms. It is doubtful whether in the human subject a pure septicemia is produced by it.

In *osteomyelitis* this organism is almost invariably present, either alone or associated with other species.

In diseases of the respiratory passages infection by the golden staphylococcus is very common. It is one of the organisms present in tonsillitis and peritonsillar abscess. We find it in the purulent infiltration of the lung which terminates some cases of pneumonia. Here it is a secondary infective agent. Not all persons are equally susceptible to staphylococcus infection. Diabetic individuals are prone to abscesses.

Almost all the *lower animals* are susceptible to natural and experimental infection by the staphylococcus pyogenes aureus. Rubbed into the skin or injected into the subcutaneous tissues of rabbits, mice, and guinea-pigs, there follows an abscess similar to local infections in the human subject from the same cause. These results are not constant, owing to variations in virulence of the micro-organism. Intravenous injection is almost uniformly followed by death in a few days. Postmortem examination reveals small abscesses in the skeletal muscles, diaphragm, myocardium, and the kidneys. The coccus, if present in the blood at all, is in very small numbers. The influence of injury upon the production of abscesses and other suppurations is very interesting and important. Thus it has been shown that quantities of a bouillon culture, which can be destroyed in the healthy body, are capable of setting up purulent inflammation at the site of some injury to the tissues.

Toxic Products. — The marked chemotaxis and the rapid necrosis of tissue produced by the staphylococcus leave little doubt that toxins of some kind are produced. Two classes of toxalbumins have been isolated from fluid cultures, one of which, insoluble in water, produces tissue necrosis and increases susceptibility; the other, soluble, is capable of conferring immunity.

Immunity Experiments.—By the injection of gradually increasing quantities, either of living or sterilized cultures, into animals, a certain amount of immunity can be brought about. Passive immunity has not been produced.

(*b*) STAPHYLOCOCCUS PYOGENES ALBUS (WHITE STAPHYLO-
COCCUS).

This micro-organism frequently occurs in pus, either alone
or associated with other forms. In morphological characters
and growth on culture media it differs from the aureus
only in the absence of pigment production.

The staphylococcus pyogenes albus from different sources
varies greatly in virulence, but on the whole it may be said
that in this respect it is equal to the golden coccus. The two
micro-organisms vary somewhat as regards the susceptibility
of different animals. In the horse infection by the aureus
is more common. In the pig we encounter the albus more
frequently. In the latter animal a septicemia, due to infection by the white staphylococcus, is not uncommon. Under
the name *staphylococcus epidermis albus* Welch has described a
variety of the staphylococcus pyogenes albus which is almost
invariably present on the surface and deep in the human epidermis. There is no good reason for considering it as other
than an attenuated form of staphylococcus pyogenes albus.

(*c*) STAPHYLOCOCCUS PYOGENES CITREUS.

This micrococcus, which in artificial culture produces a
lemon-yellow pigment, is of rare occurrence in suppurative
processes of various kinds. In size and mode of grouping it
is the same as the forms described. It ordinarily possesses
but slight virulence.

(*d*) STREPTOCOCCUS PYOGENES SEU ERYSIPELATIS.

The streptococcus pyogenes is a parasitic organism, not
finding outside the bodies of animals conditions favorable to
development. It is not an uncommon parasite upon exposed
mucous membranes, as the mouth, vagina, etc.

In the past much discussion has been indulged in as to the
identity of the streptococcus pyogenes discovered by Rosen-

bach in pus and the streptococcus erysipelatis isolated by Fehleisen from cases of erysipelas. Not less than six different varieties of the streptococcus have been described. Each presents slight variations from the typical form, but these are not sufficient to warrant a classification into distinct species. They should be considered as streptococci modified by the environments under which they have been placed.

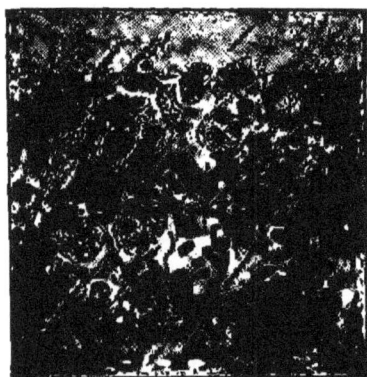

Fig. 11.
Streptococcus and staphylococcus
in pus.

Morphology.—A small coccus, 0.4 μ to 1 μ in diameter, without voluntary motion. The characteristic of this organism is the formation of chains containing from eight to twenty members or more. The streptococcus from different sources varies somewhat in size and in the length of the chains. The longest chains are obtainable by cultivation in bouillon. Sometimes we find a few elements in a chain larger than the others; this is especially marked in old cultures.

Growth on Culture Media. — *On gelatin* very small, punctiform, translucent colonies are formed, which on the surface are about 0.5 mm. in diameter. They show no tendency to increase in size. Gelatin is not liquefied. Microscopically the colonies are slightly yellow, with a regular contour.

Upon agar plates they are larger and more opaque.

Gelatin Puncture Culture.—A very delicate growth, composed of round, isolated colonies.

Upon the surface of oblique agar a thin, transparent, or slightly opaque layer develops.

Upon blood-serum the growth is similar to that on agar.

The streptococcus pyogenes does not grow upon *potato*.

In *bouillon* the character of growth varies. There may result a diffuse opacity of the medium; a flocculent precipitate is sometimes produced, the supernatent fluid remaining clear.

Milk is coagulated.

In a few days there is a diminished alkalinity in the culture media, or the reaction may become acid. The streptococcus is able to grow in slightly acid media, on which it remains alive and retains its virulence for a greater length of time. It grows with or without oxygen.

The streptococcus pyogenes is capable of multiplication between 16° and 40° C., but 37° C. is the most favorable temperature. It is killed by exposure for ten minutes to 52° to 54° C.

Staining.—With the simple dyes, and by Gram's method.

Method of Isolation.—Puncture a bleb at the edge of an erysipelatous area, and from the fluid make agar plates. The micro-organism may also be isolated by preparing plates from pus containing it.

Pathogenesis.—The streptococcus pyogenes from different sources varies greatly in virulence, often possessing little or no pathogenic action upon lower animals, inoculations as ordinarily practiced not being followed by symptoms. White mice and rabbits are most commonly used for experimental purposes. If a virulent culture be injected into the ear of a white rabbit, there may follow a local infection with transitory redness and swelling; a phlegmonous inflammation with abscess formation; a general infection without any marked local symptoms.

Virulence can be increased by passage through rabbits; in this way it is also possible to restore lost pathogenic action. A streptococcus very deadly to rabbits, after being passed several times through mice, will increase in virulence for that animal, but is no longer harmful to rabbits.

Fehleisen, in 1883, isolated this coccus from the skin of erysipelas cases and produced a typical attack by injecting pure cultures into the skin of men. Natural infection takes place in erysipelas through lesions of the cuticle or mucous membranes. Sections of skin, especially at the edge of the inflamed area, show the coccus in the lymph channels.

Uncomplicated streptococcus infection shows little or none of the tendency to localization and the destruction of tissue which characterizes staphylococcal invasion.

In *puerperal septicemia* the streptococcus pyogenes is the most common infecting organism. In the great majority of cases infection is incident to the introduction of the germ by the nurse or accoucheur a short time before or during parturition. Inasmuch as an attenuated streptococcus is often found in the vagina of healthy women, it is possible that in the lochial discharge the germ may regain its lost virulence, and that self-infection may occur.

In *scarlet fever* the occurrence of an exudation upon the tonsils, in which the streptococcus is almost invariably present, has led some to assert that it is the etiological factor in that disease. Undoubtedly it is the cause of tonsillitis in scarlet fever.

It has been shown that the streptococcus in the throat is capable of producing pseudo-membranous inflammation which presents the clinical features of diphtheria.

It is one of the micro-organisms which have been found in *ulcerative endocarditis.*

In *empyema* it occurs alone or in association.

In certain conditions which have other bacteria as their cause, secondary infection by the streptococcus is very common. This is the case in diphtheria, and adds much to the gravity of the disease. It is commonly met with in bronchopneumonia which occurs during the course of diphtheria. After death from diphtheria it has been found in the internal organs.

As a secondary infecting organism it plays a very important part in the production of the *hectic fever* which so often occurs in tuberculosis.

In *middle-ear inflammations* excited by the streptococcus, general septicemia and joint involvement have been known to follow.

Immunity Experiments.—Nothing is known in regard to the nature of the poisonous products of the streptococcus. No antitoxic substance has been found in the blood-serum of persons who have recovered from infection. Marmorek, by the subcutaneous injection of highly virulent living cultures into horses, succeeded in making them tolerant of enormous quantities of the germ. The serum of these animals is offered as a therapeutic agent in streptococcus infections, and has been used extensively in erysipelas, puerperal septicemia, and mixed infections. The method of preparation does not rest upon a firm scientific basis, and very little can be said in favor of the serum. In erysipelas it does not seem to limit the extension of the disease, nor has it any great immunizing property. Immunity has been brought about in small animals by the injection of streptococcus cultures heated to 110° C.

Influenced by the observation that a spontaneous attack of erysipelas sometimes checks the growth of malignant tumors, the streptococcus has been injected into sarcomata and carcinomata with the hope of amelioration or cure. Early investigators used living cultures, inducing thereby an attack of erysipelas. Later it was shown that as much could be accomplished by the injection of the filtrate of sterilized mixed cultures of the streptococcus and bacillus prodigiosus. The method had enthusiastic followers, and favorable results have been reported in inoperable malignant growths, particularly in sarcomata.

(*e*) MICROCOCCUS TETRAGENUS.

First observed by Gaffky in the walls and contents of tubercular cavities. It is common in the expectoration of consumptives, and is occasionally found in the saliva of healthy persons. It is not known to lead a saprophytic existence.

Morphology.—The individual coccus is spherical, non-motile, and rather large, about 1 μ in diameter. In cultures upon artificial media it has no characteristic arrangement. In tissues of inoculated animals and in sputum groups of four cells (and sometimes a single cell) are seen inclosed in a transparent capsule, the outline of each coccus being distinct and spherical. This mode of grouping, which occurs only in the animal body, has given to the coccus its name, tetragenus.

Fig. 12.
Micrococcus Tetragenus.
Spleen × 600. Flügge.

Growth on Culture Media takes place at room temperature, but that of the incubator is most favorable.

Upon gelatin plates colonies appear after thirty-six or forty-eight hours, but do not attain full development until the third or fourth day. They are round or oval, slightly elevated, and of a shining, whitish color. Microscopically they appear yellowish-brown, with smooth edges, and are made up of numerous finely granular masses.

In gelatin puncture culture growth is most abundant upon the surface. Along the needle track, round, white, dense colonies are formed, which may be distinct or more or less confluent. Gelatin is not liquefied.

Upon agar plates the colonies are small, round, and almost transparent. They have a moist, glistening appearance.

Stab cultures in agar form, upon the surface, a thick, white coat; development takes place as a uniform growth along the line of puncture.

The *growth upon blood-serum* is moist and white, without any distinctive characters.

Potato.—A thick, white, pasty growth which can be drawn out in long threads.

The coccus is facultative anaerobic.

Staining.—It stains readily with the simple watery solutions of the dyes, and also by Gram's method, which is remarkably well adapted to staining it in tissue. By the latter method beautiful specimens can sometimes be obtained from the walls of tuberculous cavities.

Pathogenesis.—*Lower Animals.*—House-mice, rats, field-mice, rabbits, and dogs are immune. White mice are peculiarly susceptible. The introduction of a minute quantity of pure culture, sputum, or saliva containing the coccus is usually followed by fatal septicemia. Virulence can be increased by rapid passage through white mice. In guinea-pigs subcutaneous injections may be followed by abscess formation with recovery, or by general infection and death. Introduced into the peritoneal cavity it sets up general purulent peritonitis.

Man.—The germ is slightly pathogenic to man. It has been found in abscesses, and is responsible in a measure for the purulent character of the inflammation in the walls of tuberculous cavities.

(ƒ) BACILLUS PYOCYANEUS.

The bacillus pyocyaneus (bacillus of blue pus; bacillus of green pus; bacterium aeruginosum) was isolated by Gessard from wound discharges which had colored the dressings green.

Besides its frequent presence in green and blue pus, it has been found iu the serous discharge from healthy wounds.

Recent observations lead us to believe that it inhabits the skin, especially about the anal region and axilla, of about 50 per cent of healthy persons, and that it is not an uncommon saprophyte in water and in the human intestine.

Morphology.— A short and slender motile rod, with rounded ends, occurring singly and in chains of four to six members. It contains cilia, which can be stained by Loeffler's method.

Growth on Culture Media.—The bacillus grows rapidly, both at room temperature and in the incubator. It is semi-anaerobic, and forms no spores.

Upon *gelatin plates* colonies are visible in twenty-four hours. Those on the surface are flat, with irregular edges. The medium between the colonies assumes a greenish hue. Gelatin begins to soften after forty-eight to seventy-two hours, and is completely liquefied on the fourth or fifth day. "Under the microscope colonies in the depth of the medium present a roundish, coarsely granular appearance, with serrated borders, and are of a yellowish-green, shining hue. The superficial ones form delicate lamina, with a smooth central depression, of a finely granulated texture, distinctly greenish in the center, but paler toward the edges."

In gelatin stab culture a funnel-shaped liquefaction occurs. The medium assumes a greenish hue, which is most marked upon the surface.

Upon oblique agar-agar the growth is dry, sometimes of a shining, greenish color. Sometimes it is white or greenish white. The medium assumes a green or bluish-green color, which, as the growth becomes older, turns brown.

Blood-serum is liquefied with the production of a greenish, and, later, a brown discoloration.

Upon potato a yellowish-green or sometimes a brown growth is developed.

Bouillon (2 per cent peptone) is turned green; occasionally a culture is met with giving a blue color.

Milk is coagulated.

In media containing albumin the bacillus pyocyaneus produces a fluorescent-green pigment. A blue pigment which is soluble in chloroform is formed in media containing peptone.

The micro-organism produces indol.

Staining.—The bacillus pyocyaneus stains with the ordinary aniline dyes, but not by Gram's method.

Pathogenic Action.—*Lower Animals.*—Inoculation of about 1 c.cm. of a virulent bouillon culture into the subcutaneous tissue of rabbits and guinea-pigs is followed by septicemia and death in from twenty-four to forty-eight hours. There is usually an abscess with much inflammatory reaction at the point of inoculation. Injections into the peritoneal cavity of rabbits and guinea-pigs induces fatal fibrinous peritonitis. The bacillus can be cultivated from all the organs.

Passage through rabbits increases virulence of the bacillus.

Feeding with toxins does not cause symptoms, while intravenous injections are followed by enteritis.

Man.—In children a fatal septicemia sometimes develops, the infection atrium being the intestinal tract.

It is sometimes present in diarrhea, and imparts a green color to the discharges.

Immunity Experiments.—Inoculation of rabbits with small, non-fatal doses is followed by immunity to lethal quantities of the germ.

The very interesting observation has been made, that animals immunized against the bacillus pyocyaneus are protected against the bacillus of typhoid fever as well. These immunized animals are also protected against infection by the bacillus anthracis.

Bacillus Coli Communis, although almost strictly a pyogenic micro-organism, will, owing to its morphological similarity to the typhoid bacillus, be considered with it.

(g) MICROCOCCUS OF GONORRHEA.

The micrococcus of gonorrhea (diplococcus of Neisser; gonococcus), discovered by Neisser (1879) in the pus of gonorrhea, is a strict parasite, being found only in the purulent discharges of that disease. It usually occurs in heaps in or upon the pus cells, and exceptionally in epithelial cells.

Fig. 13.

Pus cell containing gonococci.

Morphology.—A diplococcus, the contiguous surfaces of which are flattened and separated by a narrow cleft; the single cell varies in length from 0.8 μ to 1.6 μ; in diameter from 0.6 μ to 0.8 μ.

Growth on Culture Media.—The gonococcus does not grow upon gelatin, agar, or the liquid media ordinarily employed.

Bumm succeeded in cultivating it upon coagulated *human blood-serum* at 30° to 34° C. The growth appears as a delicate, smooth film 1 to 2 mm. wide, which has a yellowish-gray color.

Subsequently Wertheim found that a combination of human blood-serum with nutrient agar is more favorable for the growth than simple serum. Colonies on the surface are moist and grayish-white. The organism has been cultivated on pericardial and pleural exudations. Hydrocele and ascitic fluids are not as well adapted. Wertheim's mixture is particularly useful in isolating the coccus.

Acid urine to which one per cent peptone has been added is a suitable medium, the presence of urea favoring growth.

Alkalinity has a detrimental effect. The coccus grows best in a slightly acid medium, and also retains its virulence for a greater length of time.

Methods of Isolation.—According to Wertheim, several drops of gonorrheal pus are stirred into liquid human blood-serum. The serum is kept at a temperature of 40° to 45° C., and is then mixed with equal parts of liquefied agar, cooled to the same temperature. Plates are poured of the mixture, and are kept in the incubator. Colonies appear in forty-eight to seventy-two hours, and can be transferred to a tube containing blood-serum-agar mixture.

A simple method of isolation which will also be found useful is the following:

Human blood-serum is coagulated and sterilized in a Petri dish and the pus distributed over the surface in minute clumps. If a case has been selected in which the gonococcus is abundant, with but few other species present, a pure culture can very often be started.

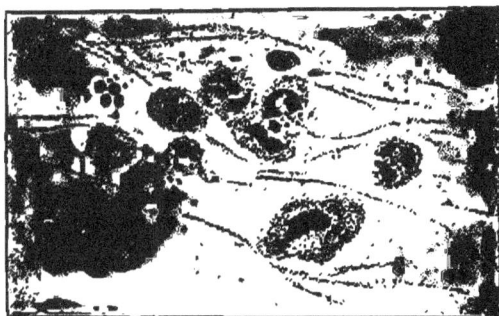

Fig. 14. Gonococcus in pus.

The minimum temperature at which development takes place is 32° C., the most favorable being 36° to 37° C.

Vitality.—The gonococcus from different sources varies in the length of time vitality is retained upon artificial media. As a rule it can no longer be transplanted after five or six days, although some observers have seen it grow when transplanted after an interval of one month. It can be maintained in artificial culture a greater length of time if isolated early in the course of acute gonorrhea.

According to some investigators gonorrheal pus preserved on bits of cloth may remain virulent two and one half months.

Staining. — The gonococcus can be stained with dilute aqueous solutions of the ordinary aniline dyes, but not by Gram's method. Double staining can be accomplished by treating the cover-glass preparation first by Gram's method and then with a dilute aqueous solution of Bismarck brown. The gonococcus not retaining the violet is stained brown. Beautiful specimens can be obtained by double staining with methylene blue and eosin.

Pathogenesis.— *Man.*— The gonococcus upon mucous membranes produces a suppurative inflammation. In adults surfaces lined with squamous epithelium, as the mouth and vagina, are infrequently attacked. The parts most commonly affected are the male urethra, cervix uteri, endometrium, fallopian tubes, vagina in children, conjunctiva, and the peritoneum. In peritonitis, which sometimes follows acute gonorrhea in the female, the gonococcus has been found in the peritoneal cavity at the autopsy.

The coccus enters the body, but not the nucleus, of the pus cell. It passes between the epithelial cells to the subepithelial tissue, and from here it may get into the circulation. In this way gonorrheal endocarditis, arthritis, and iritis have been explained. The inflammation produced by this micro-organism in mucous membranes is of a purely exudative type, without any tendency to the formation of granulation tissue. In gonorrheal arthritis, however, the inflammation is of the plastic type.

Cases of suppurative inguinal adenitis and prostatic abscess occur in the course of gonorrhea. This may be due to the gonococcus, or to secondary infection by other pyogenic bacteria.

The gonococcus may remain in the semi-latent infections of the genito-urinary tract for an indefinite length of time and retain, in large measure, its pristine virulence. Subcutaneous injections are not followed by abscess formation.

Lower Animals.—Mice and young guinea-pigs can be infected by intraperitoneal injection of pure cultures of the gonococcus, particularly if some of the culture medium is introduced, or if there be extensive injury to the peritoneum. Rabbits and rats are immune.

Immunity Experiments.—Nothing is known of the toxic products of the gonococcus. One attack of gonorrhea does not confer immunity. There seems to be great variation in the susceptibility of different individuals to the disease. It is greatest in infants, children, and young adults.

BACTERIA OF CROUPOUS PNEUMONIA.

Croupous pneumonia is an inflammatory condition of the lung, during the course of which one or two micro-organisms are always present.

(a) MICROCOCCUS OF CROUPOUS PNEUMONIA.

Synonyms: Micrococcus Pasteuri; micrococcus of sputum septicemia; diplococcus lanceolatus; diplococcus pneumoniæ (Fraenkel-Weichselbaum); pneumococcus.

Discovered by Sternberg (1880) in the blood of rabbits inoculated subcutaneously with human saliva, and by Pasteur in the same year in the blood of rabbits which had been inoculated with the saliva of a child dead of hydrophobia. In 1883 Talamon demonstrated its presence in the sputum of croupous pneumonia. In 1886 Fraenkel and Weichselbaum published independent researches, showing this micrococcus to be present in the exudate of seventy-five per cent of all cases of croupous pneumonia, and pointing out its etiological relation to the disease. It can be demonstrated in the saliva of most healthy persons.

Morphology.—The pneumococcus in sputum, pneumonic lung, and in the blood of animals occurs in pairs (diplococcus). The elements of the pair are oval or lancet-shaped, and are united by their rounded extremities. Short chains are sometimes seen, but these as a rule are found to be made up of united diplococci. In the animal body, and occasionally in the initial growth upon blood-serum, the pneumococcus is surrounded by a distinct capsule. The capsule is not found in the germ grown upon alkaline artificial media. The elongated and encapsulated cells seen in the animal body represent the highest phase of development in the pneumo-

coccus.   In artificial culture media almost spherical forms
are produced, and chains of six or eight elements are not
uncommon.

Growth on Culture Media.—The pneumococcus is a
strictly parasitic organism.   Its cultivation outside the body
is comparatively difficult.   Culture media in which this
micrococcus is sown must have an alkaline reaction, the
slightest acidity being sufficient to prevent development.

On gelatin plates, at
22° to 24° C., it grows
very slowly; often not
at all.   On the third
or fourth day the colo-
nies appear as minute
white or bluish-white
points which, under
the microscope, are
sharply defined and
granular.

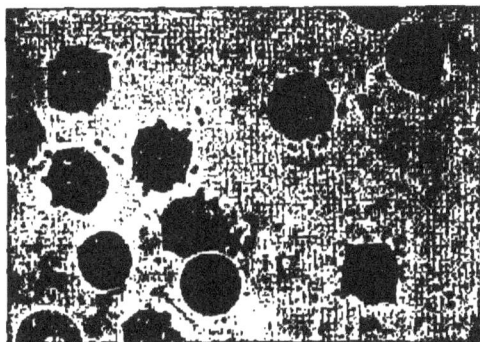

In gelatin puncture
cultures the growth is

Fig. 15.   Pneumococcus in blood of rabbit.
x 1000.   Fraenkel and Pfeiffer.

very much like that of the streptococcus pyogenes.   Small,
white, isolated colonies develop along the entire needle track.
Gelatin is not liquefied.

Upon agar plates, at 35° C., the colonies appear on the
second day as small, slightly elevated, transparent drops,
resembling under the microscope the colonies on gelatin.

The coccus produces *upon oblique agar and blood-serum* a
thin, transparent film.   Opacity of the growth denotes contamination.

In bouillon a slight opacity is produced.

Infusion of rabbit flesh is more suitable than the bouillon
prepared in the ordinary way.

Milk is coagulated by the virulent germ, but not by
attenuated varieties.

No growth is observed upon *potato.*

The most favorable temperature for the growth of the diplococcus of pneumonia is 35° to 38° C.; the extremes at which multiplication takes place, 18° to 42° C. It is facultative anaerobic.

Vitality.—Upon artificial media the pneumococcus will not retain its vitality unless frequently transferred to fresh nutrient material, and, even though this is done, not rarely it fails to grow after a few transplantations. Agar and gelatin cultures, after a week or ten days, are usually found to be sterile. In bouillon, especially if the acidity produced is neutralized, somewhat older cultures can be obtained.

Staining.—The pneumococcus stains well with the simple aniline dyes. For demonstration in tissue Gram's or Weigert's method is applicable.

For staining the capsule Friedlaender suggested the following: The exudate, dried and fixed in the flame, is flooded with glacial acetic acid, which is replaced at once, without washing in water, by Ehrlich's aniline-water-gentian-violet solution in sufficient quantity to remove the acetic acid. After two or three minutes wash in 2 per cent sodium chloride and examine in the same. The bacillus is deeply stained, while the capsule has a lighter color.

Pathogenesis.—The virulence of the pneumococcus is sometimes great, sometimes slight. It is most pronounced in the rusty sputum of pneumonia, in the recently solidified lung, and in the blood of infected animals after death. Its pathogenic property is quickly lost by cultivation.

Lower Animals.—The pneumococcus is pathogenic for mice, rabbits, and, in a lesser degree, for guinea-pigs. Chickens and pigeons are immune. In dogs, sheep, and cats abscess follows subcutaneous injection of virulent cultures. Owing to the mutable virulence of this micro-organism, the results of experiments upon animals are not uniform. There may follow septicemia, with or without inflammation at the point of inoculation. If a less virulent culture has been used an abscess follows. Rabbits are most commonly used for experiments. In this animal 0.1 to 0.2 c.cm. of a bouillon culture suffices to produce death in twenty-four to forty-eight hours. A little of the rusty sputum of pneumonia or material obtained by scraping a consolidated pneumonic lung produces the same result. Croupous pneumonia has been produced by the injection of pure cultures into the lung and pleural cavity of the rabbit, dog, and sheep.

Man.—The development of the pneumococcus in the human subject is accompanied by fibrinous or suppurative inflammation, depending upon the structure of the tissue and the virulence of the germ. The lungs and meninges of the brain are most often the seat of pneumococcus invasion. In croupous pneumonia its constant presence in the exudate, its high virulence at the outset, and gradual attenuation as the disease progresses toward crisis, place its etiological relation to this disease beyond question. Recent observations have disclosed the fact that in pneumonia this micro-organism often invades the blood in small numbers, and that infection of the kidney is common. It is also the cause of pericarditis, endocarditis, and menigitis, which sometimes develop during the course of pneumonia. In some cases of pneumonia a true septicemia occurs, and the micrococcus can be cultivated from the blood and all internal organs.

The pneumococcus was found in twenty-seven out of forty-five cases of *purulent meningitis* examined by Netter.

In *tonsillitis* and in *purulent otitis media* it is also frequently met with.

In *purulent inflammation of the parotids*, in *arthritis and empyema following pneumonia*, it has been found in pure culture.

Cases of *peritonitis, ovarian and mammary abscesses* due to infection by the pneumococcus have been reported.

Toxins — Immunity Experiments. — An albumose has been isolated from the blood of infected animals and from pure culture of the germ upon artificial media. Intrathoracic injection of the albumose from these sources was followed by dyspnea and elevation of temperature. At the dissection the lung was found consolidated, the pleura inflamed.

It sometimes happens from lack of virulence that fatal results do not follow experimental infections. Such animals have been found immune against new infection. A short-lived immunity in rabbits has also been produced by inoculation with filtered toxins. Experiments are not convincing that the blood-serum of artificially or naturally immune animals has a decided protective or curative influence.

(*b*) BACILLUS PNEUMONIÆ. (FRIEDLAENDER.)

The bacillus was isolated by Friedlaender, in 1883, from a pneumonic lung. Subsequent researches have shown that it is occasionally present on the healthy mucous membrane of the mouth and respiratory tract, and also in the air.

Morphology. — The bacillus of Friedlaender (pneumobacillus) is a short, non-motile rod, $0.75\,\mu$ wide and 1 to 2 μ long; the ends are rounded. It usually occurs singly, sometimes in pairs and in chains of three or four cells. Longer filaments are met with, both in the tissues and in artificial culture. Each cell, in the tissues, is surrounded by a thick capsule. In our alkaline culture fluids this capsule does not appear.

The bacillus does not form spores.

Growth on Culture Media. — Friedlaender's bacillus grows luxuriantly in *gelatin* at room temperature. Large colonies are visible upon the *plates* in twenty-four to forty-eight hours, and appear to the naked eye as round, white, shining masses with elevated centers.

In gelatin puncture culture development takes place from top to bottom of the medium. Over a small area of the surface a white, elevated, button-like mass develops, which has a smooth, shining appearance. Gas bubbles appear in the medium. A brownish discoloration takes place in older cultures.

Upon agar growth is also abundant, and resembles that on gelatin.

Upon blood-serum it is similar and slightly viscid.

It produces *upon potato* an abundant "yellowish-white and very thick film, in which gas bubbles develop at incubator temperature."

Bouillon becomes opaque, and a mucoid film forms upon the surface.

Milk is not coagulated.

It does not produce indol.

Friedlaender's bacillus grows between 16° C. and 40° C., both in the presence and absence of oxygen.

Vitality is retained for a year on the surface of agar or in gelatin. It is destroyed by a temperature of 56° C.

Staining.—The bacillus colors with the simple aniline dyes, but does not retain the stain in Gram's method.

The capsule can be stained as recommended by Friedlaender.

Pathogenesis.—*Lower Animals.*—In the experiments of Friedlaender the bacillus was injected into the thoracic cavity of dogs, rabbits, guinea-pigs, and mice. It proved to be pathogenic for dogs, mice, and guinea-pigs, in which it produced a purulent pleuritis and areas of pneumonia. Rabbits were found to be immune.

In man Friedlaender's bacillus is found in a small per cent of cases of lobar pneumonia—"nine times in one hundred and twenty-nine examined by Weichselbaum, three times in seventy cases examined by Wolf." We have demonstrated it by culture in one case in the organs of an infant two months of age.

BACILLUS OF INFLUENZA.

The constant presence of a certain bacterium during the course of influenza allows the inference that it is an infection due to a specific germ.

Bacillus of Influenza is the name given to this germ by R. Pfeiffer, who was the first to isolate it. It is always present in the sputum of influenza patients, and sometimes invades the blood.

Morphology.—A very short, plump, non-motile rod ($0.2\,\mu$ –$0.3\,\mu \times 0.5\,\mu$). It is often found in pairs.

Sporulation does not take place.

Growth on Culture Media.—No growth takes place on the ordinary media. Fresh blood (red or white corpuscles) is essential to its development. For cultivating the bacillus fresh blood is spread in a thin layer over the surface of agar or blood-serum.

The colonies appear as small, transparent, colorless beads which show no tendency to coalesce. Microscopically they appear homogeneous. Old cultures are slightly yellow.

The bacillus will not grow below 28° C.

Vitality.—The bacillus of influenza shows little resistance to desiccation. In the dry state, at 20° C., it expires in twenty-four hours. It also dies rapidly in distilled water. In bouillon it lives several weeks at 20° C. Cultures on fresh blood retain their vitality a long time if transferred to a fresh tube every three to five days. Exposure to 60° C. for five minutes is fatal.

Method of Isolation.—The suspected material is spread over the surface of agar or blood-serum, which is covered by a thin layer of sterile blood. If the material containing the bacillus furnishes the proper nutriment, as is the case in blood, pus, and sputum, it can be spread directly upon the agar or blood-serum.

Staining.—The influenza bacillus does not stain readily with the simple dyes. Loeffler's alkaline methylene blue penetrates better. It does not stain with Gram's method.

Czenzynski's solution has been used for staining the bacillus in blood and sputum. (See chapter on staining.)

Pathogenesis.—*Lower Animals.*—Rubbing pure cultures of the germ into the mucous membrane of the nose, or injections into the peritoneal cavity of rabbits and monkeys, bring about fever of several days' duration. This is probably a toxic action. The bacillus has

Fig. 16. Influenza bacillus. Sputum. Flügge.

also pyogenic properties, as abscess formation following its subcutaneous injection has shown.

Man.—In influenza patients the nasal and bronchial secretions constantly contain the bacillus, at times in immense numbers. It is especially abundant in the green, purulent bronchial discharges, in which it can usually be demonstrated by direct microscopic (stained specimen) examination.

In severe cases of influenza a lobular inflammation may take place in the lung. The bronchi and alveoli become filled with pus. Many of the pus cells are filled with the bacilli.

The bacilli, as a rule, do not enter the blood and viscera.

Immunity Experiments have not been satisfactory. One attack confers no immunity against subsequent ones.

MICROCOCCUS OF EPIDEMIC CEREBRO-SPINAL MENINGITIS.

Diplococcus Intracellularis Meningitidis. — Weichselbaum gave this name to the germ isolated by him in 1887 from the exudation of a number of cases of epidemic cerebro-spinal meningitis. He believed to have discovered in it the specific organism for this disease. It is found in the leucocytes of the exudate, but also in tissue cells.

Morphology.—The bacterium is a diplococcus with flattened adjacent sides, like the gonococcus. The dimensions are somewhat greater than in the latter.

Growth on Culture Media.—*Gelatin*—No growth.

On agar or glycerin agar plates the deep colonies are very delicate, almost invisible. Under the low power of the microscope they appear slightly granular.

The superficial colonies are larger. They appear as flat discs, almost transparent at the periphery. Towards the center they become thicker and are slightly pigmented brown. The colonies appear in about twenty-four hours.

On blood-serum a thin, almost invisible, slightly granular layer is formed.

Vitality.—Pure cultures lose their vitality in six days. If transferred to a fresh medium every two days they will remain alive for a long time.

Method of Isolation.—By inoculating surface of glycerin agar with meningeal exudation; also by the plate method.

Staining.—It stains readily with the simple aniline dyes. Results with Gram's method have been contradictory.

For tissue, Loeffler's methylene blue is the best stain.

Pathogenesis.—*Lower Animals.*—The results of subcutaneous inoculation have all been negative. Intraperitoneal injections in mice produced death in thirty-six to forty-eight hours. The diplococcus was found in the spleen and in the pleural and peritoneal exudations at the autopsy. The diplococci were nearly all in the pus cells. Inoculation into the meninges caused inflammation and punctate hemorrhages.

Man.—The coccus has been frequently found in the healthy nose. This has led to the supposition that infection of the meninges takes place through the nasal cavity, its accessory sinuses, and the middle ear.

The presence of the bacterium in the exudation of epidemic cerebro-spinal meningitis and its position in the pus cell makes its specific character likely.

BACILLUS OF TUBERCULOSIS.

Tuberculosis is an acute or chronic infectious disease, produced by a specific germ, the bacillus tuberculosis, and characterized by distinct and constant pathological changes. It is frequent in man and most lower animals.

The *bacillus tuberculosis* was discovered by Koch in 1882.

It is a strict parasite, and is found only in the lesions of tuberculosis and in discharges therefrom.

Morphology.— A non-motile rod with rounded ends, varying in width from 0.2 μ to 0.4 μ, and in length from 1.5 μ to 4 μ.

It is, as a rule, slightly curved or bent on itself. Although found singly, two to six cells are not infrequently seen clinging together end to end or associated in irregular clumps.

In stained specimens the bacillus frequently presents a beaded appearance not unlike a chain of cocci. These are poorly nourished cells, and can always be seen in old cultures. The cause of this appearance is not known.

The bacillus from different animals often presents slight variations, morphological as well as biological.

Spores have never been found.

Growth on Culture Media.— As the bacillus grows very slowly, it requires special environments and extreme care in the technique of isolation and culture.

Glycerin-agar (five per cent glycerin) is the medium most favorable for its growth.

Blood-serum is especially useful in starting cultures from the tissues of animals. They will develop on either of these media in two to four weeks, if kept at body heat, appearing either as small, dry, grayish-white scales above the surface or as irregular, mealy clumps. The growth can readily be lifted from the medium.

The scales and clumps are made up of tangled masses of bacilli. The cells are held together by an exceedingly tenacious substance, and it is with considerable difficulty that they are broken apart. As the growth continues the scales may bunch into heaps.

Bouillon, to which glycerin has been added, offers a suitable medium for growing the bacillus. The bouillon remains clear, but a granular sediment is formed while a mycoderma develops on the surface.

Potatoes, moist macaroni, and radishes have also been successfully used as media for culture.

A temperature of over 30° C. is essential to the growth of the tubercle bacillus. Development is most abundant between 37° and 39° C.

Vitality.—The dried bacillus will withstand a temperature of 100° C. for several hours. In the moist state it dies in four hours when exposed to 55° C.; in one minute when exposed to 95° C. Freezing has no effect.

It shows great resistance to desiccation, being able to live for several months at 35° C. in the dry state.

The resistance to chemicals is also marked, especially in sputum, where the mucus prevents the chemical from coming in contact with the cell. Bichloride of mercury is least effective in sputum. Carbolic acid, five per cent, and absolute alcohol are better.

Putrefaction of tuberculous material has little effect upon the vitality of the tubercle bacillus.

If transferred every six weeks to fresh culture medium, vitality and pathogenic property are retained for years.

The detrimental influence of sunlight upon Koch's bacillus is deserving of particular notice. In diffuse light a culture will die in five or six days; in direct rays of the sun, in a few minutes to several hours.

Methods of Isolation.— Isolation of the tubercle bacillus is not an easy task. In consequence of its slow growth,

any other species that should chance to be present would
soon overgrow it.   If a post-mortem examination can not
be obtained upon the body of a tuberculous subject, the best
method of isolation is to inject tuberculous sputum into a
susceptible animal (guinea-pig).   If the animal does not
succumb to the action of other bacteria contained in the
injected material (streptococci, etc.), tuberculosis will develop
in from three to six weeks.   After infection is well established the animal is killed, and material from the interior

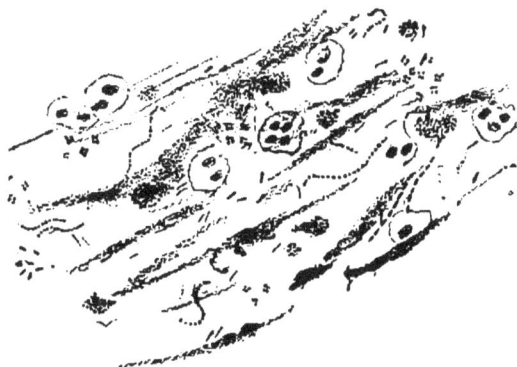

Fig. 17.   Tubercle bacillus in sputum.

of a tuberculous lymphatic gland is, under strict aseptic
precautions, broken down and spread over the surface of
oblique agar or blood-serum.   It should be rubbed into the
medium.   The bacillus, for the first few generations, grows
best upon blood-serum.   To avoid drying of the medium
while in the incubator, the tubes should be corked or capped
with rubber, the cotton having previously been singed and
moistened with a carbolic acid solution to prevent contamination by moulds.

Staining.—The ordinary staining solutions act only after
prolonged exposure of the bacillus to them, or by the addition
of a mordant.   On the other hand the bacillus gives up the
dye, when once it has taken it on, most reluctantly.   These
properties enable us to distinguish the tubercle from other

bacilli by microscopical examination, and upon them the various methods of staining the micro-organism are based. The peculiar reaction of the tubercle bacillus to stains is due to certain fats contained in it. If these be removed by treating the bacillus with caustic potash its specific staining characters are lost.

The bacillus is readily *stained in sputum* and other material containing it, which for this purpose is spread in a thin layer on a cover-slip or slide, dried and fixed in the flame.

NOTE.—The germs are most numerous in the little, white, cheesy masses often seen in sputum of tuberculous subjects.

Of the methods for staining sputum, Guenther's and Gabbet's have given the best results.

GUENTHER'S METHOD.

1. Float slip with fixed material, spread side down, on Ehrlich's solution of fuchsin* in a watch-glass. (See chapter on stains.)
2. Heat until it gives off vapor, and allow it to rest.........................1 to 5 minutes.
3. Move cover-glass about actively in a 3 per cent solution of hydrochloric acid in alcohol ...1 minute.
4. Rinse thoroughly in water.
5. Counter-stain with very weak, watery solution of methylene blue ....1 to 2 minutes.
6. Rinse in water.
7. Allow it to dry in the air.
8. Pass through flame several times to fix the stain.
9. Mount in balsam.

Those bacteria which have retained the red stain are tubercle bacilli. All other bacteria and the tissue elements appear blue. Where we find a number of uniform rods which have taken the fuchsin stain, the diagnosis of tuberculosis can be made with certainty.

_____

*On account of its stability Ziehl's is frequently substituted for Ehrlich's solution in this step of the method.

The presence of one or two rod-like bodies which have retained the red color would not warrant the diagnosis, as cholesterine crystals, fragments of hair, crystals of fuchsin, etc., are often seen in the field, and may be mistaken for bacilli.

Fraenkel has suggested the combination of the counter-stain (methylene blue) with the bleaching agent. The method, as it has been modified by Gabbet, is rapid of execution, but not so reliable as Guenther's. In the decolorization by the strong acid, in Gabbet's method, the bacilli, if present in small numbers, may not retain the stain and be overlooked.

GABBET'S METHOD.

1. Stain with Ziehl's solution (cold)....5 to 15 minutes.
2. Gabbet's solution, according to thickness of the specimen........................$\frac{1}{2}$ to 1 minute.
3. Wash in water.
4. Dry and mount.

This method is not adapted to making permanent specimens, as the bacteria in time lose their color.

The bacillus in pure culture may be examined by the same methods.

Staining Tubercle Bacillus in Tissue. — Small pieces of tissue are hardened in absolute alcohol and sections prepared in the usual way.

The most useful method for staining tuberculous tissue is:

THE KOCH-EHRLICH METHOD.

1. Keep sections in cold Ehrlich's solution for 12 hours. (If warmed, $\frac{1}{2}$ hour will be sufficient.)
2. Wash in 25 per cent watery solution of nitric acid for several seconds.
3. Wash in 60 per cent alcohol.........1 to 3 minutes.
4. Counter-stain in weak, watery solution of methylene blue or Bismarck brown (the former where fuchsin was used in Ehrlich's solution, and the latter where gentian violet has been used) ....1 to 5 minutes.
5. Pass rapidly through 60 per cent alcohol.
6. Dehydrate in absolute alcohol.......30 to 60 seconds.
7. Clear in oil of origanum.
8. Mount in balsam.

Gram's Method answers for staining the bacillus of tuberculosis in tissue and on the cover-slip, but is not a good differ-
ential test, as it reacts similarly with a large number of
bacteria.

Pathogenesis.—*Lower Animals.*—The tubercle bacillus is
pathogenic to nearly all the lower animals. Infection experi-
mentally induced is always fatal in guinea-pigs after a period
varying from six weeks to six months or longer. Spontaneous infection is observed in rabbits, monkeys, cattle, pigs,
and cats. Horses, dogs, mice, rats, and canary birds are least
susceptible. Sparrows and the cold-blooded animals are
immune.

The action of the bacillus can best be studied by injecting
a small quantity of a pure culture or a bit of tuberculous
sputum into the peritoneal cavity, anterior chamber of the
eye, or under the skin of a rabbit or guinea-pig. A proliferative inflammation takes place which in two to four weeks
results in the formation of numerous small nodes.

The presence of the bacillus causes, first, proliferation of
the fixed tissue cells, which are large, with prominent nuclei,
and are spoken of as epithelioid cells. Multiplication by
division takes place. The bacilli lie in and between the
cells. Not infrequently a nuclear division takes place without division of the cell body which results in large multi-
nuclear cells (giant cells). Along with this proliferation
inflammation takes place, accompanied by marked invasion
of leucocytes.

When the circumscribed areas have become fully invaded
by leucocytes the central portion of the tubercle undergoes
coagulation necrosis. The nuclei first disappear, and then
the cells break down and form a cheesy mass. The mass
may become encapsulated by a fibrous wall and the bacilli
remain inert, or the process may extend, involving the
walls of blood-vessels, and the bacilli may be carried away
by the circulation to other parts of the body and set up

the same localized changes, or miliary tuberculosis may develop.

The peculiar action of the tubercle bacillus is due to chemical products of its growth. Wyssokowitsch, Vissman, and others were able to bring about the formation of tuberculous nodules without necrosis by intravenous injection of dead cultures of the bacillus.

Along with the local changes there is always more or less elevation of temperature, emaciation, etc. Secondarily, degeneration of the parenchymatous organs often takes place.

Man.—The lung is the most frequent seat of tubercular lesions in the human adult. The respiratory tract is, in by far the greatest number of cases, the avenue by which the bacillus enters the body. The sputum of tuberculous individuals becomes dried and pulverized, and when inhaled with dust forms the most common medium of infection both in the human subject and lower animals. The bacilli may cause local pathological conditions in the lung, or they may be carried to remote organs, particularly the lymphatic glands, and in the latter may remain latent for an indefinite length of time.

The bacilli also enter the body by way of the alimentary canal. They are swallowed with food—the infected milk or meat of tuberculous cattle—or with saliva, contamination of which may occur through bacilli floating in the air. Auto-infection of individuals with pulmonary tuberculosis results from swallowing saliva contaminated with the pulmonary excretions. In the alimentary canal the bacilli usually invade the lymph-adenoid structures and lead to the formation of nodules and finally ulceration in the solitary and agminated glands. Infection by the alimentary canal is common in children.

Exceptionally tubercle bacilli enter the system by the skin (abrasions). Lupus is most likely the result of an infection of this sort.

The bacilli may invade the body of the child in utero, remaining latent for a variable length of time (several months). The frequency with which it is found without lesions, in the bodies of infants, leads to the belief that congenital infection very often takes place.

The pathological changes in man differ on the whole but little from those in lower animals. From the primary foci the bacilli are sometimes carried, by the blood or lymph, to remote parts, viz., liver, spleen, larynx, where the same characteristic changes take place. A general miliary tuberculosis results if the bacilli enter the blood-vessels in sufficient numbers.

Primary tubercular lesions of the joints, bones, glands, etc., are accounted for by the entrance of bacilli through the mucous membrane of the alimentary or respiratory tract. Passing the points of entrance without causing a lesion, they are carried to remote or more susceptible parts.

Mixed infection of the tubercle bacillus with the pyogenic bacteria or the pneumococcus is not infrequent. The prognosis of these cases is unfavorable.

Toxic Products.—Many attempts have been made to isolate from the tubercle bacillus the specific products upon which its pathogenic action depends. The studies of De Schweinitz upon the chemical composition of the bacillus and its products are of great value. He has "obtained a substance corresponding to a nucleo-albumin which appears to be the fever-producing principle of the germ." From ordinary culture media—but in larger quantity from a medium containing acid potassium phosphate, ammonium phosphate, asparagin, and glycerin—he succeeded in isolating "a crystalline substance, soluble in water, . . . having the formula $C_7H_{10}O_4$, which is the formula of teraconic acid, an unsaturated acid of the fatty series."

This substance, in doses of 0.015 gram subcutaneously, caused in healthy guinea-pigs a hemorrhagic exudate and disintegration of the tissue about the point of injection. De Schweinitz concludes that this is a specific secretion of the germ, and that it is responsible for the necrosis of tissue in tuberculosis.

Tuberculin.—In 1890 Koch announced the discovery that by the injection of a glycerin extract of the tubercle bacillus (tuberculin) protection could be secured and the life of animals afflicted with tuberculosis prolonged.

The method of preparing tuberculin, according to Koch, is as follows: The tubercle bacillus is grown in alkaline bouillon containing four per cent of glycerin. At the end of six to eight weeks in the incubator at 37° C. development ceases. The culture fluid is now sterilized by heat, filtered through porcelain, and evaporated in vacuo to one tenth its original volume. This is the ordinary tuberculin. In doses of 0.1 c.cm. it causes a rise of temperature in tuberculous guinea-pigs. Much smaller doses — 0.001 c.cm. — cause elevation of temperature and other constitutional symptoms in the human subject affected with tuberculosis.

Tuberculous animals cease to react to tuberculin after the second or third injection.

Tuberculin in large doses sets up an acute inflammatory process in the vicinity of the tubercular areas. It does so rapidly, and in that way cuts off the nutrition and brings about disintegration of the tuberculous mass. In this way superficial tubercular conditions, as lupus, may break down and be exfoliated. Where the process is in the internal organs it has been shown that the use of tuberculin in large doses is dangerous. It not only breaks down active tubercular processes, but also the old encapsulated masses in which the bacilli are in a latent condition. The bacilli may enter the blood-vessels and set up general miliary tuberculosis. On this account the use of tuberculin has been almost abandoned in therapeutics. As a characteristic elevation of temperature follows its injection into tuberculous bodies and not in healthy individuals (unless the dose be very large), the injections are still made in cattle for diagnostic purposes, and are of great value.

Koch has recently made the announcement of a new tuberculin. The dried cultures of highly virulent bacilli are triturated in an agate mortar until the bacilli have entirely disappeared, then mixed with distilled water and sedimented with a centrifuge. After half an hour in the centrifuge the upper part of the fluid is thin, transparent, slightly opalescent, and gives all the reactions of the old tuberculin, while the lower part is thick and adheres to the bottom of the tube. The sediment is dried and the process of grinding and sedimenting repeated until no deposit remains.

The fluid obtained by the second and subsequent manipulations is called by Koch TR. With this substance he has been able to produce in healthy guinea-pigs immunity to the virulent tubercle bacillus, and to cure infection that was not far advanced. A few results of its use have been published, showing it to be curative in skin tuberculosis (lupus). Koch is very sanguine of its success as a therapeutic agent early in the course of tuberculosis.

Immunity Experiments.—In the guinea-pig immunity to tuberculosis has been established by subcutaneous injection of attenuated cultures of the living germ, and following this, when recovery has taken place, with virulent bacilli. There can at present be no question that an antitoxic substance exists in the blood of animals so treated, and that in larger animals systematically injected with the living or dead germ this substance can be produced in quantity sufficient to make it available for the treatment of tuberculosis in the human subject. Still, it must be said that we can not with antitoxic serum cure experimental tuberculosis with the same uniformity that is seen in the antitoxic treatment of diphtheria and tetanus.

By the injection of increasing quantities of sterilized

cultures, filtered and unfiltered, of the tubercle bacillus into horses and goats, several observers have obtained a serum for which is claimed a curative influence in human tuberculosis not too far advanced.

The bacillus of fowl tuberculosis (bacillus tuberculosis gallinarium ; bacillus tuberculosis avium) shows striking similarity to Koch's bacillus as regards its size, shape, and reaction to stains. Distinct peculiarities of growth and pathological action class it as a separate species.

Growth.— The colonies on *blood-serum* or *glycerin-agar* develop more rapidly (eight days at body temperature) than those of tuberculosis. They are white and have a moist or waxy appearance. The surface is smooth and the colonies show no tendency to heap up, as the dry colonies of tuberculosis. They are not so tenacious, being readily broken down with a platinum needle. Occasionally they are pigmented.

The bacillus develops best between 35° and 45° C. (tuberculosis between 30° and 40° C.). It resists high temperature longer than the tubercle bacillus. It does not sporulate.

Staining.—The bacillus stains in the same manner as Koch's bacillus, but is more readily decolorized.

Pathogenesis.—Fowls are the most susceptible to natural and experimental infection, miliary tubercles and rapid emaciation following the injection of pure cultures into a vein or the peritoneal cavity. Feeding experiments give negative results. Injection into mammals (guinea-pigs, rabbits, etc.) as a rule sets up only a localized inflammation. It may, however, lead to generalized disease and death.

NOTE.—The action of tubercle bacilli is just the opposite. It is pathogenic for mammals and non-pathogenic for fowls.

The pathological changes differ slightly from those of tuberculosis. The nodes are made up principally of fixed tissue cells. Giant cells are only exceptionally found. Infiltration of leucocytes is slight, and caseation of the center of the nodes infrequent. The disease shows no predilection for the lungs. The most extensive changes take place in the liver.

Infection by the bacillus of fowl tuberculosis has been observed in man.

BACILLUS OF LEPROSY.

Leprosy is due to infection by a specific germ, the bacillus of leprosy, which is found only in proliferations of that disease.

Morphologically it differs from the bacillus of tuberculosis in that it is shorter, and not so frequently bent or curved.

Growth.—All efforts at isolation have proven futile.

Staining.—It takes the simple stain and Gram's. It shows the same tendency to retain dyes as the tubercle bacillus. Saturated specimens are acted upon slowly by acids and alcohol.

Pathogenesis.—*Lower animals* are practically immune.

In man the bacilli are present in great numbers throughout all parts affected with leprosy. The nodules which are formed are made up of granulation tissue. They do not undergo caseation, but frequently ulcerate. The bacilli are found in the tissue cells. Typical leprosy followed the experimental inoculation of leprous tissue into a condemned criminal. Death occurred five years later.

SYPHILIS.

Although syphilis is looked upon as an infectious disease, and bacteria have been found in syphilitic lesions, experiments upon man and animals in support of their relation to the disease are wanting.

Bacillus of Syphilis is the name given by Lustgarten to a micro-organism found by him in syphilitic tissue.

Morphology. — The bacillus in appearance is almost identical with that of tuberculosis. It is somewhat longer and usually more bent (irregularly), and often shows a thickening of one end or the other. It appears solitary or arranged in heaps in syphilitic lesions, in and between the tissue cells.

The bacterium has never been cultivated.

Staining.—The stained bacillus withstands the action of mineral acids and alcohol, not to the extent, however, of the tubercle bacillus. Lustgarten stained sections with Ehrlich's solution, rinsed in alcohol and bleached in permanganate of potash.

Pathogenesis.—Inoculation of lower animals with syphilitic tissue has always failed to produce specific lesions. Whether the bacillus described by Lustgarten is the cause of the disease has not been positively demonstrated. Its pathological significance has particularly been questioned, since a bacillus alike in appearance and in reaction to stains has been found in smegma, between the labia minora, in the anus, etc.

This, the **Smegma Bacillus,** is said to be a different germ by most authors. Although about the size of the micro-organism observed in syphilitic tissue, its dimensions are more variable. It does not take on stain readily, and resists the acids longer than the former. The bacillus has not been obtained in pure culture. It acts negatively when injected into animals.

BACILLUS OF GLANDERS.

Glanders is an acute or chronic infectious disease of animals of the equine species, characterized by the formation of new growths which show a speedy tendency to ulcerate. The initial lesion of glanders in the horse is commonly upon the Schneiderian mucous membrane (nasal glanders). In the more chronic forms there is enlargement of subcutaneous lymphatic glands (farcy).

Bacillus mallei, the specific germ of glanders, was isolated by Loeffler in 1882. It is strictly a parasite, and is found only in the discharges of the diseased parts of animals affected with glanders, and is most abundant in recent lesions.

Morphology.—It is a non-motile rod, having about the length of the tubercle bacillus, but almost twice its thickness. The ends are more or less rounded. It occurs singly in the lesions of glanders, but in cultures, especially upon potato, several cells are often seen united. The cell protoplasm exhibits the same irregularities that were noted in the bacillus tuberculosis, and have been attributed to a degenerative change in the protoplasm.

Spores are not formed.

Growth in Culture Media.—The glanders bacillus is facultative anaerobic. It can be cultivated at temperatures between 25° and 42° C., but best at 37° C. Glycerin-agar and blood-serum are the most favorable media.

Upon agar or glycerin-agar plates the colonies appear on the second or third day as white or yellowish-white, spherical bodies, which microscopically are yellowish-brown, slightly granular, with smooth edges. The colonies are not characteristic.

A stroke culture on this medium appears on the third day as a whitish, almost transparent film, which attains its full development on the fifth or sixth day.

Upon blood-serum colonies appear on the third day; they are yellowish-white, and finally coalesce to form a thick growth. The medium is not liquefied.

In bouillon a diffuse opacity is produced. The reaction becomes acid.

The growth upon potato is characteristic. It appears first as an amber-yellow covering, which rapidly increases in thickness and assumes a darker tint. At the end of a week the culture has a reddish-brown color.

Vitality.—In the moist state the bacillus is destroyed by exposure to 70° C. for six hours. A temperature of 90° to 100° C. destroys it in three minutes.

Corrosive sublimate is the most effective disinfectant salt. It destroys the bacillus in ten minutes in a 1–5,000 solution.

The micro-organism can not resist desiccation under the conditions of experiments for a longer period than ten days at 25° C., but in the discharges of glandered animals clinical experience points out a greater period of viability; upon artificial media cultures can ordinarily be kept alive for three months without transferring.

The glanders bacillus does not find outside the body conditions suitable for its multiplication.

Method of Isolation.—The plan in general use for diagnostic purposes is to introduce the suspected material into the peritoneal cavity of a male guinea-pig, at the same time crushing the testicle. If the disease be glanders the bacillus localizes itself in the injured part and suppuration occurs. Death follows in from twelve to fifteen days. Involvement of the testicle can be noted on the second or third day, and when the disease has advanced somewhat further the skin and tunica vaginalis may be incised and the glanders bacillus obtained in pure culture in the usual way.

This method has been adopted because the preponderance of other species makes its isolation from the nasal secretion almost impossible.

Staining of the bacillus in pure culture can be accomplished by the simple method, though it takes the stain reluctantly. Loeffler's alkaline methylene blue is best. It does not stain with Gram's method.

It is difficult to demonstrate the bacillus in tissue, owing to the fact that the color imparted to it is very quickly lost when treated with decolorizing agents.

The best method is that of Kuehne. (See chapter on staining.) In all cases the sections should be made from the fresh nodules, for in these the bacillus is most abundant.

Pathogenesis.—Glanders is a disease of horses, mules, and asses, but may be contracted by other animals. The cat tribe is susceptible; rabbits, sheep, and dogs are almost immune; swine, cattle, white mice, rats, and house-mice completely so. The guinea-pig is especially susceptible to inoculation, and is the most common victim of experiment. A very remarkable susceptibility to the disease is found in field-mice. In these animals subcutaneous injection of a virulent pure culture is followed by death from septicemia within three or four days. Scattered throughout the internal organs are numerous grayish points scarcely visible to the naked eye, which have a structure very similar to that of the young miliary tubercle, differing from it only in the absence of the central necrosis characteristic of the latter. In guinea-pigs tumefaction follows in four or five days; an ulcer forms at the point of inoculation, the neighboring lymphatics enlarge, and symptoms of general infection develop. Death usually occurs at the end of four or five weeks. At the post-mortem grayish or yellow masses, resembling very closely old tuberculous nodules, are found in all the internal organs.

In man infection occurs chiefly among hostlers and others who come in contact with diseased animals. The pathological changes differ little from those in lower animals.

Mallein.—The filtered products of bouillon cultures of the glanders bacillus are used as a diagnostic agent under the name of mallein.

The technique of preparation is as follows: The virulent glanders bacillus is sown in alkaline bouillon, to which 4 per cent of glycerin has been added. At the end of four to six weeks the culture fluid is sterilized by heat and filtered through unglazed porcelain in order to remove the bacilli. The sterile filtrate is then evaporated in vacuo. The degree of concentration of this filtrate (mallein) varies in the different preparations of mallein on the market. The effect of its introduction into a glandered horse is an elevation of temperature not less than 1.8° C., and as a rule not more than 3° C. The elevation of temperature occurring is accompanied by other symptoms of illness and the appearance of a swelling at the point of injection. No such effects are produced in healthy animals.

BACILLUS OF RHINOSCLEROMA.

"Rhinoscleroma is a chronic affection of the skin, and especially of the mucous membrane of the nares, characterized by the formation of tubercular thickenings of the skin and tumefaction of the nasal mucous membrane, followed sometimes by ulceration. It prevails in Italy, Austria, and to a slight extent in some parts of Germany."

The bacillus of rhinoscleroma, which has been discovered and isolated from the diseased areas, is a small rod, with rounded ends, occurring singly; occasionally in pairs. Sometimes the cell is almost spherical, and resembles very closely a coccus. It is surrounded by a capsule both in the body and upon culture media. It is non-motile.

Spores are not developed.

Growth on Culture Media resembles that of the bacillus of Friedlaender, but is decidedly viscid upon agar.

Staining is accomplished both with the simple colors and by Gram's method.

Pathogenesis.—The bacillus does not give rise in lower animals to the same character of lesion in which it is discovered in man. It is pathogenic for rabbits, mice, and guinea-pigs, the action being similar to that of Friedlaender's bacillus.

BACILLUS OF TYPHOID FEVER.

The bacillus of typhoid fever (Bacillus of Eberth-Gaffky) was discovered by Eberth in 1880–81 in stained sections of the organs of a typhoid cadaver. Three years later it was obtained in pure culture by Gaffky. It is found in the feces, urine, blood, and some of the viscera of typhoid patients. The bacillus is capable of leading a saprophytic life.

Fig. 18.
Typhoid bacillus, showing flagella.

Morphology.—A short rod, 0.5 μ–0.8 μ × 1 μ–3 μ, with rounded corners. It has active motion, due to the presence of ten to eighteen flagella, which are distributed along its periphery. In artificial culture the bacillus sometimes forms long threads. When growing in the body or on nutrient agar it is plumper than when cultured on gelatin or potatoes.

Spore formation does not take place.

Growth on Culture Media.—*Upon gelatin plates* the superficial colonies are most characteristic. After thirty-six to forty-eight hours they are irregularly oval, or round, and transparent, and show under the low power of the microscope a homogeneous surface, marked with numerous furrows. They have been compared to grape leaves. Old colonies are slightly granular and have a brown pigmentation in the center. Deep colonies have sharply defined edges. They are granular and pigmented brown. Gelatin is not liquefied by the growth.

In gelatin puncture culture a delicate layer with serrated edges forms on the surface at the point of puncture and gradually spreads to the walls of the tube. It is whitish and semi-transparent. Along the line of puncture a finely granular growth takes place, which in old cultures is brown.

On agar a gray, translucent growth develops along the track of the needle.

On blood-serum a white line develops along the needle track, extending but little beyond it.

In bouillon a diffuse clouding takes place without a trace of gas production.

The *growth on potatoes* usually covers the surface as a transparent, thin, tenacious coating, which is seen only by pulling at it with a platinum needle. This growth, which characterizes the typhoid bacillus, takes place only upon acid potatoes. If the potato reacts neutral or alkaline the culture often remains limited to the line of inoculation; sometimes it is pigmented.

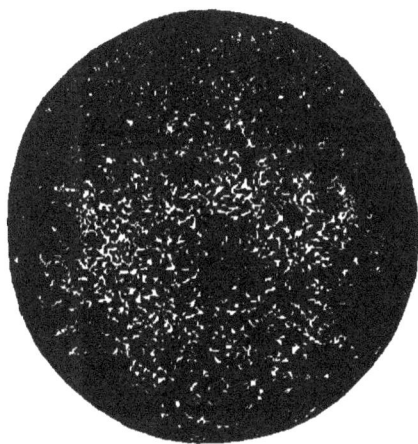

Fig. 19.

Colony of typhoid bacillus three days old. × 100. Fraenkel and Pfeiffer.

On Holz's medium colonies have a characteristic transparent, shining appearance. This medium is favorable to the growth of the typhoid bacillus, whereas a number of saprophytic organisms do not prosper upon it.

On Elsner's medium the growth is similar.

Milk is a favorable medium, and is not changed by the presence of the typhoid bacillus.

The bacillus grows between 9° C. and 42° C.; best at 37° C.

Other characteristics of the growth of the typhoid bacillus are:

(1) Its slight acid production.

(2) That it does not cause fermentation of glucose.

(3) That media, free of sugar, become highly alkaline.

(4) That it does not form indol (in contradistinction to bacillus coli communis).

Vitality.—Exposure of the bacillus to 60° C. for one half to one hour destroys it. Cold has no effect.

It retains its vitality for three months in distilled water.

Experiments show that the typhoid bacillus will remain alive for a greater length of time in water containing a small amount of organic matter and but few saprophytic species. In grossly contaminated water it is soon crowded out by the numerous saprophytic forms, but according to several observers putrefaction has slight influence upon the typhoid bacillus. "Added to putrefying feces it may preserve its vitality for several months, in typhoid stools for three months, and in earth upon which bouillon cultures have been poured, five and one-half months."

Desiccation kills it in five to twenty days.

It shows a marked resistance to carbolic acid, being able to prosper in the presence of 0.1 to 0.2 per cent solution.

Pure cultures retain their vitality for many months.

Methods of Isolation.—To obtain the bacillus from alvine dejections of a typhoid patient is rather difficult. It can not be discovered until about the end of the first week of the disease. The object of the various methods is to secure, by the addition of chemical agents, a medium which will restrain the growth of the numerous saprophytic forms without restricting the typhoid bacillus. The ability of the typhoid bacillus to grow in weak solutions of carbolic acid is taken advantage of in the following method : Add 0.5 per cent carbolic acid to liquid gelatin, into which some of the fecal matter of a typhoid patient has been stirred. It is allowed to congeal on plates. The presence of the carbolic acid prevents the development of the mass of saprophytes. The bacillus typhosis and the bacillus coli communis usually develop in greater number than other species, and can be separated from them and from each other.

For methods of detecting the typhoid bacillus in water, see chapter on Examination of Water.

Staining.—It is more difficult to stain the typhoid bacillus than most bacteria, as it does not take the simple dyes readily and bleaches with Gram's method.

Sections of tissue are best stained with Loeffler's alkaline methylene blue solution, twelve hours or more, washed in water, dehydrated in alcohol, cleared in oil, and mounted.

For demonstrating flagella, see chapter on this subject. Stained bacilli often (especially from potato culture) show darker spots at the poles, and sometimes vacuoles at different parts of the cell. This is due most likely to retrograde changes. They have been mistaken for spores.

Toxic Products.—Brieger found in cultures of the typhoid bacillus a specific ptomaine, *typho-toxin*, having the formula $C_7H_{17}NO_2$, which produced in guinea-pigs salivation, diarrhea, frequent respiration, dilatation of the pupil, and death. Brieger and Fraenkel isolated a toxalbumin from cultures in blood-serum and bouillon.

Pathogenesis.—*Lower Animals.*—All attempts to reproduce typhoid fever in lower animals have failed. Taken by the mouth or inhaled, the germs are inert; injected into the peritoneal cavity death ensues as the result of a toxic action. Sometimes, where large quantities have been used, they multiply and set up inflammation.

Man.—The germs always enter the human system by the alimentary canal, which they reach with food and water, or with the saliva contaminated by hands or utensils. Several epidemics have occurred as a result of contamination of milk by water used in cleansing the cans or diluting the milk. The bacillus shows a predilection for the solitary glands and Peyer's patches. It is also found in the mesenteric glands, spleen, liver, and kidneys. The distribution of the typhoid bacilli in the tissues is peculiar. They are extra-vascular, scattered about in small, irregular clumps, being most abundant in the spleen and in lymphatics leading from ulcerated patches.

In the solitary and agminated glands a proliferation of the tissue cells takes place with infiltration of leucocytes and finally *necrosis*. Necrosis seldom takes place in the viscera. During the course of or after typhoid fever the bacillus sometimes induces an inflammatory process in remote parts which terminates in suppuration. It has frequently been found alone in these secondary abscesses. Such complications are, however, usually due to infection with streptococcus, staphylococcus, pneumococcus, etc. The typhoid bacillus probably enters the blood in small numbers. In this way any organ may be invaded. This also explains the infection of the fetus in utero.

Sometimes the bacillus is taken up so rapidly that no changes take place in the intestinal glands.

Immunity Experiments. — As in other diseases, one individual is more susceptible to typhoid fever than another.

The injection of gradually increasing quantities of the bacillus, dead or living, into lower animals brings about immunity, and their serum takes on protective and curative properties. Humans who have recovered from an attack of typhoid are subsequently immune in a high degree. Numerous experiments have been made with their serum as a protective and curative agent. None have been successful.

Serum Diagnosis. — As a result of infection by the typhoid bacillus the albuminous fluids of the body acquire the property of causing rapid loss of motion in the typhoid bacillus and its aggregation in clusters. The exact nature of the substance which causes this reaction is not understood, but it is in combination with, or in close relation to, globulin and fibrinogen; for, if these substances be precipitated, the agglutinating property of the serum is lost. It has been shown that this reaction is acquired by the serum in a large majority of cases as early as the fifth or sixth day of the disease.

Method of Making the Test.— As originally suggested by Widal, two methods were used. For demonstrating it in the test-tube he recommended that one part of blood or blood-serum be added to a young culture of the typhoid bacillus in alkaline bouillon. The reaction consists in the precipitation of the typhoid bacilli as a flocculent sediment, and is complete in three or four hours. A more delicate test, and that largely used at present, is made by adding the blood-serum or the serum obtained from a blister in the proportion of one part to thirty parts of a twenty-four-hour-old bouillon culture of the typhoid bacillus in the hanging drop and examining with a magnification of about five hundred diameters. Cessation of motility and clumping often occurs before the instrument can be focused, or it may be delayed as much as six hours, depending upon the agglutinating strength of the serum, the character of the bacillus, and the dilution. If the reaction does not take place in two hours with a dilution of one part of blood to thirty parts of culture fluid the test should be considered negative. The agglutinative property of dried blood is preserved indefinitely, and is therefore very largely used by practitioners for transmitting specimens to the laboratory. Instead of a bouillon culture a small quantity of the germ from the surface of nutrient agar is distributed in distilled water upon the cover-glass and the serum added to this.

Principal diagnostic points of the typhoid bacillus are its shape, motility, the invisible pellicle it forms on potato; further, that it does not produce indol or fermentation of glucose, and that milk is not coagulated. Its slight acid production is also characteristic. To these can be added the reaction of the bacillus when mixed with the serum of a typhoid patient.

BACILLUS COLI COMMUNIS.

Synonyms: Bacterium coli commune (Escherich); colon bacillus; motile feces bacillus; bacillus pyogenes fetidus (Passet); bacillus Neapolitanus (Emmerich).

Possibly no other pathogenic bacterium is so widely distributed. It is always found in the intestine of man and lower animals under normal and pathological conditions. Being able to lead a saprophytic existence, it is frequently encountered outside the body in soil, water, air, milk, etc.

Morphology.—A short rod with rounded ends, resembling the typhoid bacillus. It occurs singly, in pairs, and may grow out into long threads. It contains flagella, but these are not so numerous as in the typhoid bacillus. Virulent bacilli, in the first few generations after isolation from the body, are slightly motile, but in subsequent cultures motility is lost.

Growth on Culture Media.—The growth is more rapid than the typhoid bacillus, otherwise shows similarity on solid media.

On gelatin plates and puncture cultures the colonies are almost identical with the typhoid colonies. They are tinged somewhat with yellow, and frequently are oval or "whetstone" shaped.

On agar and blood-serum the growth is more luxuriant than the typhoid, and more opaque on account of greater pigmentation.

Bouillon.—A milky opacity of the medium is produced in eighteen hours at incubator temperature.

Potato.—A luxuriant, shining growth of a dirty brown color.

Milk is coagulated in one to two days.

Other characteristics of the growth of the colon bacillus are:

1. The production of indol.
2. The fermentation of glucose.
3. Marked acid production.

The vital resisting power of the bacillus coli is greater than that of the Eberth bacillus, and if the two are present primarily in equal proportion in the medium the former soon takes the precedence.

The colon bacillus also grows with more facility in the acid media used in the isolation of the typhoid bacillus than the latter; upon a plate containing both species the colon bacillus is recognizable by larger size of the colonies.

Staining.—It stains with the simple aniline colors, but is bleached with Gram's method.

Pathogenesis.—The bacillus is pathogenic for most *lower animals.* Fatal spontaneous infection occurs in pigs, calves, and rabbits, the most prominent symptom being diarrhea. Virulent cultures injected into the peritoneal cavity of mice, rabbits, or guinea-pigs multiply in the body and rapidly enter the circulation, causing septicemia and death. Less virulent cultures multiply in the body, but do not set up a fatal septicemia. Death occurs after a variable length of time, the principal changes being necrotic areas in the liver and kidney in which the bacillus coli is present. An acute enteritis is always set up. Subcutaneous injections often, especially in rabbits, produce abscesses. Given by the mouth the bacillus is inert unless it is extremely virulent. The disease-producing property varies greatly in cultures from different sources.

Man.—In the human the principal pathogenic action of the bacillus is its *pus production.* As a rule large quantities of the bacillus coli are present in the normal bowel. Under unfavorable conditions of the body and increased virulence of the germ they may set up an *infectious enteritis.* In *infection of the biliary passage* and *abscess of the liver* the bacillus coli is frequently found in pure culture. It is present in nearly all cases of peritonitis following gunshot wounds and other injuries of the intestine. It has also been found in subcutaneous suppurations, puerperal infections, urethritis, cystitis, etc.

The principal diagnostic points of the colon bacillus are its size, shape, slight motility or its absence, more luxuriant growth on all media than typhoid, milky opacity it gives to bouillon, and its characteristic dirty, shining growth on potatoes; furthermore, its production of indol, marked fermentation of glucose, acid production (coagulation of milk), and its negative reaction to the typhoid serum. It is more pathogenic to lower animals than typhoid, and less pathogenic to man.

SPIRILLUM OF ASIATIC CHOLERA.

The spirillum of cholera (comma bacillus of Koch; cholera vibrio) was discovered in 1883 by Koch. He was able to isolate it in all cases from the dejecta of cholera patients. The blood and internal organs never contain it. It is not found in healthy individuals nor in any other disease.

Fig. 20. Cholera Spirillum.
× 600. Koch.

Morphology.—In the discharges of cholera victims and upon the surface of solid culture media the organism is a short, curved rod (vibrio), varying in length from 0.8 μ to 2 μ. The thickness is one fifth to one fourth the length. The curve is more marked in the young cells. Under adverse influences of nutrition, such as low temperature, presence of antiseptics, and exhaustion of the nutrition in the medium, the cells often grow out into spiral forms. (Fig. 20, *a*.) They show the same tendency when grown in bouillon. The cholera vibrio has a single long flagellum on one end, and has active voluntary motion.

Spore formation has not been observed in Koch's vibrio.

Growth on Culture Media. — *Gelatin Plates:* After twenty-four hours, at 22° C., small white dots appear, which, under the microscope (low power), are yellowish-white, glistening, and have an irregular edge. They are slightly granular, becoming more so as they get older. Their center also gets darker. The gelatin is slowly liquefied and funnel-shaped depressions formed, at the bottom of which lie the white, button-like colonies.

Gelatin Puncture Cultures.—In twenty-four to forty-eight hours a white opacity has formed along the line of puncture, which in time becomes more dense. The gelatin around it is slowly liquefied, especially near the surface, giving to the area of liquefaction a funnel shape, which gradually widens and deepens until at the end of six days it reaches the wall of the tube. The liquefaction extends downward and involves the entire gelatin in fourteen days.

Agar.— A grayish - yellow, translucent growth takes place along the line of inoculation. It has nothing characteristic.

Blood-Serum. —Liquefaction of the medium occurs. The growth has no distinguishing features.

Bouillon.—A luxuriant growth, causing cloudiness of the medium in twelve to twenty-four hours (at 37° C.). At about the same time a delicate white mycoderma, made up almost entirely of cholera spirilla, forms on the surface. This pellicle becomes thicker, and finally breaks up and sinks to the bottom.

Fig. 21.
Cholera Spirillum.
Gelatin culture 60 hours old.
Shakespeare.

Potatoes.—No growth takes place at room temperature; at 30° to 35° C. a light brown film.

Milk.—The growth is luxuriant, without perceptibly changing the medium.

Another biological characteristic of the cholera bacillus is the production of *indol.* Nitrites are also formed. The test for indol in this instance is made by the simple addition of sulphuric acid, which gives a red color. The addition of a nitrite is superfluous, as it is formed during the cell growth.

It has been found that this nitrose indol reaction, which was formerly looked upon as characteristic of the cholera vibrio, takes place with several other vibrio species. If the reaction does not take place, however, it excludes Koch's vibrio.

The minimum of temperature at which the micro-organism will grow is 9° C., the maximum 40° C. It grows best between 22° and 37° C.

Vitality.—The spirillum succumbs in less than an hour when exposed to 55° C. Freezing for a week has no effect. Desiccation destroys it in two to twenty-four hours.

The presence of saprophytic germs in large numbers would, after a time, be detrimental to development. At first the cholera bacillus may outgrow the other bacteria present, but as soon as the nutriment it requires is exhausted, or conditions of nutrition such as change in reaction take place, the other germs take precedence, and the cholera vibrio may in time entirely disappear. In ordinary water the germ will live a long time. In distilled water it dies for want of nutritious substances.

Acidity of the medium is detrimental, also the presence of various chemicals (sublimate 1–1,000 ; carbolic acid 2 or 3 per cent).

The presence of oxygen is not essential to the growth (facultative anaerobic).

The vibrio of cholera will live a long time in pure culture : Gelatin, three to five months; agar, six to nine months.

If left on the medium too long, so as to make the conditions of nutrition unfavorable, the cells rapidly undergo changes. They form threads, or the cells become swollen and distorted into various shapes; these are the so-called *involution forms.*

Method of Isolation.—Some of the mucus in the dejecta is stirred into liquefied gelatin, which is diluted in the customary way and poured on plates. The growth is rapid. In twenty-four hours small colonies are visible. Where the vibrio is present in proportionately small numbers this method does not answer. It is best then to bring the material into a solution of 1 per cent peptone and $\frac{1}{2}$ per cent chloride of sodium, and put the same into the incubator. In six hours the cholera spirilla come to the top and form a delicate pellicle which is almost a pure culture. Some of this can be spread on gelatin and agar or inoculated into tubes for plates.

Staining.—The cholera spirillum takes the simple stains readily. It bleaches with Gram's method. The stained specimen does not show the curve of the organism as well as the unstained.

Pathogenesis.—*Lower animals* never acquire cholera by natural infection. It is difficult to produce it artificially, owing to the inability of the spirilla to resist the acidity of the gastric juice. Koch was able to bring about infection in the guinea-pig by neutralizing the acidity of the stomach just prior to feeding with cholera germs. At the same time he lessened peristalsis by the administration of opiates. Typical symptoms of cholera and death resulted. In rabbits the same result was attained by injections into the auricular vein. Metschnikoff produced it by rubbing the culture into the mucous membrane of the mouth. The pathological changes were similar to those found in man.

Man.—Numerous cases are on record in which the ingestion of cholera spirilla produced Asiatic cholera. The best known of these are the experiments of Pettenkofer and Emmerich, who swallowed the bacteria. The result in the former was simply a diarrhea, while Emmerich contracted typical Asiatic cholera. The specific germs were found in abundance in the intestinal discharges of both.

In mild cases of the disease there is simply a slight swelling and hyperemia of the mucous membrane. A large quantity of transudate containing white flocculi of mucus collects in the intestinal canal. The specific germ is present almost in pure culture at this stage. Later there is a marked change in the mucous membrane. It is covered with hemorrhagic spots, principally at the site of Peyer's patches and solitary glands, and not infrequently necrosis takes place here. The spirilla are found in the intestinal follicles between the epithelium and in the intertubular tissue, but do not penetrate below the basement membrane. Necrosis takes place most frequently in protracted cases, and the entire mucous membrane becomes dark red, especially in the lower part of the gut. The necrotic patches are covered with a pseudo-membrane. The intestinal contents are bloody and have a disagreeable odor. At this stage other bacteria predominate. The cholera germs have diminished in number. During the process in the intestine severe constitutional symptoms—fever, coma, delirium—are present, which finally end in death.

Various theories have been advanced as to the character of the disease-producing element and its mode of action. It is generally believed that intestinal ptomaines play no part in it.

Pfeiffer's theory, the most reasonable, is that a certain part of the bacterial cell is made up of toxic substances. He explains the absorption of these substances, which can only be taken up through a broken surface of the mucous membrane, in a necrosis of the epithelium following the bacterial invasion. The poison acts upon the circulation, and finally causes death. The intracellular poisons act very rapidly, the symptoms following the injection almost at once. They disappear rapidly after the last stage of the disease (in six to eight days).

Chemicals, heat of 60° C., and drying have a detrimental influence upon the poison.

Toxic Products.—A toxalbumin has been isolated by Brieger. Petri was able to isolate from culture fluids a poison similar in reaction to peptone, which was called by him *toxopeptone.* It is able to withstand steam at 100° C. Cadaverin, hutrescin, and other ptomaines are produced in albuminous fluids, but are in no way related to the symptoms of cholera.

Pfeiffer demonstrated that in the living or carefully killed cholera spirillum there exist poisonous substances which have a paralyzing action upon the circulatory apparatus and the heat-producing centers. Their action is rapid and transient. Absorption does not take place through the intact epithelium.

To obtain these cholera poisons Pfeiffer sterilized a fresh culture upon agar by exposing it to chloroform for ten minutes. Ten milligrams of the culture treated in this way was sufficient to kill a two-hundred-gram guinea-pig by intraperitoneal injection, there being present before death all the symptoms of the algid stage of cholera.

These cholera poisons are very unstable, and are quickly destroyed by exposure to 60° C. Chemical agents and drying also have a marked detrimental effect. They do not pass into the culture fluid. If the spirillum be exposed to conditions which are sufficient to destroy these primary poisons, there remain others which produce the same symptoms but are ten to twenty times less virulent.

Immunity Experiments.—It has been found that the blood-serum of a human, just over an attack of cholera, would protect a guinea-pig if injected before infection takes place. The protective property of the blood is at its height four weeks after the beginning of the disease. In two to three months or less it is lost.

Pfeiffer immunized rabbits and guinea-pigs by the injection of dead spirilla into the peritoneal cavity until they were finally able to withstand lethal doses of the living germ.

Introduction of the blood-serum of these immunized animals, diluted with one hundred parts of bouillon, to give it volume, enabled susceptible animals to survive an otherwise fatal dose of the vibrio. Pfeiffer further observed that the vibrio lost its motility, that it aggregated in clumps and speedily underwent disintegration and death. This reaction, spoken of as "Pfeiffer's phenomenon," distinguishes the cholera spirillum from certain other forms resembling it.

Haffkine practices a protective inoculation in man, using either living or dead germs by subcutaneous injection.

The bacteriological diagnosis of cholera is at times easy. Where the spirilla are abundantly present in the stool a direct examination of this, if the clinical symptoms are those of cholera, is enough to make a diagnosis. Where the germs are less abundant a thorough bacteriological study is necessary. The micro-organism must be isolated and its peculiarities of growth and pathogenesis taken into consideration.

There are a number of spirilla which show similarity as regards the morphology, biology, and pathogenesis, from which the cholera spirillum must be differentiated.

Serum Diagnosis.—It has recently been observed that the blood of persons with Asiatic cholera contains, on the first or second day of the disease, a substance which posesses a specific paralyzing action upon the cholera vibrio in artificial culture. A small quantity of blood from a suspected case is added to a motile culture in the hanging drop. The reaction consists in loss of motility and clumping together of the vibrio, a phenomenon the counterpart of that observed by Pfeiffer in the peritoneal cavity of guinea-pigs.

The normal blood-serum of the ox has a very remarkable agglutinative power upon the cholera spirillum, but it acts likewise upon other forms of spirilla.

SPIRÍLLA WHICH MAY BE CONFOUNDED WITH THE CHOLERA SPIRILLUM.

FINKLER-PRIOR SPIRILLUM.

The Finkler-Prior spirillum (vibrio proteus) is comma-shaped, but is larger than Koch's spirillum. Its body is thicker, and the ends taper. It also has a single flagellum. The micro-organism was isolated from the intestinal discharges of a case of cholera nostras.

Growth on Culture Media. — *On gelatin plates* the growth is more rapid than that of the comma bacillus, colonies being almost twice as large after twenty-four hours. The colonies are perfectly round, have a smooth edge, and are darker in color. The gelatin is more rapidly liquefied.

In gelatin puncture cultures speedy liquefaction of the medium takes place, reaching the walls of the tube at the surface in about twenty-four hours.

On agar there is nothing characteristic, a moist, whitish layer quickly forming on the surface.

Blood-serum is rapidly liquefied.

The growth upon potatoes distinguishes it from the cholera germ. In forty-eight hours at room temperature a grayish-yellow, smeary pellicle with a white edge has formed.

A strong odor is given off during the growth of the Finkler spirillum. Indol is not produced.

Vitality.—It withstands desiccation and the presence of other germs much better than Koch's spirillum.

Pathogenesis. — *Lower animals* are affected about the same as by the comma bacillus.

The germ is non-pathogenic to man.

SPIRILLUM TYROGENUM.

The spirillum tyrogenum (S. of Deneke; cheese spirillum) was isolated from cheese.

Morphology. — Comma-shaped, uniform in thickness, a little smaller than Koch's spirillum. It has a flagellum at one end, and is motile.

Growth on Culture Media. — *On gelatin plates* the growth is rapid, colonies being visible in twenty-four hours as small, round, white bodies. Under the low power of the microscope they show a smooth edge and a dark, greenish-brown center with a lighter peripheric zone. Liquefaction of gelatin begins early.

In gelatin puncture culture the medium is fluidified more rapidly than in cholera cultures, but not as speedily as in Finkler's spirillum.

Potato. — Growth takes place only at body temperature as a moist, yellowish film.

Pathogenesis. — By Koch's method of bringing about infection guinea-pigs have succumbed to its action. It is non-pathogenic to man.

METSCHNIKOFF SPIRILLUM.

Metschnikoff spirillum is a curved rod found in a fatal disease of fowls, characterized by diarrhea.

Morphology. — It is shorter and thicker than the cholera spirillum, so that it sometimes resembles a coccus. It has active motility, and bears a single flagellum at one pole.

Growth on Culture Media. — *On gelatin plates* growth is very rapid, and liquefaction of the medium more energetic than in cholera cultures. The colonies at twenty-four to thirty hours, under a low magnifying power, show a yellowish-brown, granular center. They have a peripheric, clear zone, which is marked by delicate radii which extend into a zone of liquefied gelatin.

In gelatin puncture cultures growth is twice as rapid as in cholera cultures, and liquefaction of gelatin is more pronounced. There are no other characteristics.

On potato at 37° C. a delicate brown film develops. It is smooth and glistening.

Like the cholera spirillum, the micro-organism produces indol and nitrites simultaneously.

Staining.—The vibrio stains by the simple methods, but not by Gram's.

Pathogenesis.—While the spirillum resembles that of cholera in its biological and morphological characters, it can be differentiated by its pathogenic action upon *lower animals.*

Subcutaneous injections of the bacterium kill pigeons in twenty-four hours. The tissue around the point of injection becomes swollen and discolored (yellow), and is infiltrated with a serous exudate which contains large numbers of Metschnikoff's spirillum. Guinea-pigs also succumb in twenty-four hours after subcutaneous injections. The tissue becomes swollen with a bloody, serous exudation containing many spirilla.

Infection through the mouth can only be brought about by Koch's method of neutralizing the gastric juice and stopping the peristalsis of the bowel. Animals treated in this way develop acute enteritis and die in collapse.

WATER VIBRIONES.

Besides the spirilla mentioned there are a number of other varieties, found principally in well- and river-water, which very much resemble the vibrio of cholera in one or several of its characters. Their non-pathogenic character to man would hardly justify their enumeration.

BACILLUS OF DIPHTHERIA.

Diphtheria is an infectious disease in which the specific germ (bacillus of diphtheria) remains localized in the part affected (usually throat), and the toxic products which are taken up into the blood cause more or less severe constitutional symptoms.

The bacillus of diphtheria (Klebs-Loeffler bacillus) was first recognized by Klebs (1883) in stained preparations of diphtheritic false membrane. Loeffler, in 1884, obtained it in pure culture, and by inoculation gave rise to the disease.

The Klebs-Loeffler bacillus is present in all cases of true diphtheria. It occurs in the false membrane, and especially in the older parts thereof. As a rule it does not invade the internal organs or even the tissues in the immediate vicinity of the pseudomembrane, but it has in several instances been found in the internal organs after death.

Morphology.—The diphtheria bacillus is especially characterized by variability in form. In blood-serum cultures it is most often seen as a single straight or slightly curved thick rod with rounded ends, varying in length from 2 μ to 3 μ, and in width from 0.5 μ to 0.8 μ. In unstained preparations strongly refractive granules are seen, sometimes at each pole, sometimes in other parts of the cell. These bodies color more intensely than other parts of the rod, giving rise to uneven staining. There are often three or four in a single cell, and the bacillus then resembles a chain of three or four streptococci.

Spore formation has not been observed.

Growth on Culture Media.— *Upon gelatin plates* development takes place rather slowly. The colonies attain full growth at the end of a week, but even after this time are small and not distinctive. The medium is not liquefied.

In gelatin puncture culture small round colonies are strung along the needle track, the growth resembling that of the streptococcus pyogenes.

On plates of glycerin-agar, at 35° C., colonies appear in twenty-four or forty-eight hours. They are round, or slightly oval, with irregular edges, and of a grayish-white or yellowish-

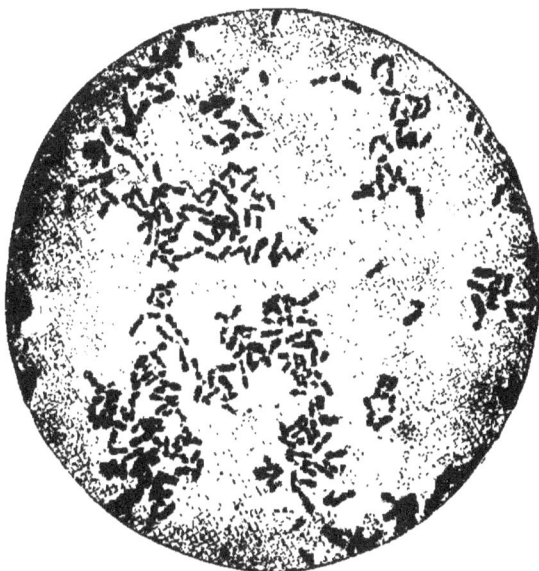

Fig. 22. Bacillus diphtheriæ. ×1000. Fraenkel and Pfeiffer.

white color. Microscopically they appear coarsely granular, with irregular, ill-defined edges.

Stroke cultures upon this medium appear as a thin white film, feeble during the first few transplantations, but increasing in vigor with each until it is abundant and covers the surface.

Loeffler's Medium.—The most favorable medium for the bacillus diphtheria is that devised by Loeffler. (See chapter on preparation of media.) At 37° C. the colonies appear in eighteen to twenty-four hours. They are small, yellowish-white, and have a glazed appearance.

Bouillon.—The bacillus grows in clumps, which may be so small as to appear to the naked eye as a diffuse cloudiness of the medium. A slight pellicle forms upon the surface after three or four weeks. The reaction of the medium, which at first changes to acid owing to fermentation of the muscle sugar, soon becomes alkaline.

Upon potato development is scanty. It occurs as an almost invisible, dry, thin, glazed film, extending but little beyond the point of inoculation.

The diphtheria bacillus requires a neutral or slightly alkaline medium. It grows with or without oxygen, and between 20° and 42° C., 35° to 38° C. being most favorable.

Vitality.—The bacillus is destroyed in the presence of moisture by exposure to 58° C. for ten minutes.

Cultures may retain their vitality for several months. When dried upon silk threads for several weeks colonies are still developed in a suitable medium. In dried diphtheritic membrane, preserved in small fragments, the bacillus retains its vitality for nine weeks, and in larger fragments for twelve to fourteen weeks. (Sternberg.)

Bacteriological Diagnosis of Diphtheria.—The morphological and cultural peculiarities of the diphtheria bacillus are sufficiently distinctive to enable us to make a rapid and certain diagnosis of the disease by bacteriological methods. Advantage is taken of the rapid growth of the bacillus upon Loeffler's serum mixture. A portion of the suspected exudate is carefully removed with a sterile cotton swab and stroked gently over the oblique surface of the medium, which is then placed in the incubator at 35° to 37° C. In from twelve to eighteen hours colonies are visible. Cover-glass preparations are now made from these, and stained in Loeffler's alkaline methylene blue solution. If it is desired to start a pure culture of the bacillus this can be done by stroking the swab over the surface of three or four serum tubes. Isolated colonies will develop on one or other of the tubes, and can be transferred as the beginning of a pure culture. Difficulties sometimes arise from improper collection of material.

The swab, which to begin with must be sterile, should not touch any other part than the false membrane. No antiseptic must be used for at least four hours before the material is collected.

Staining. — The diphtheria bacillus stains best with Loeffler's alkaline methylene blue. It is susceptible to staining by Gram's method.

Pathogenesis.—*Lower Animals.*—In the experiments of Welch and Abbot inoculation of a platinum loop of a pure culture, growing on glycerin-agar, into the trachea of a kitten caused the formation of a false membrane, similar in all respects to that seen in the same situation in man. Loeffler produced an identical process in the trachea of rabbits. The guinea-pig is acutely susceptible to the diphtheria poison, subcutaneous inoculation with 0.1 to 0.5 c.cm. of a bouillon culture resulting fatally in from one to five days. Rats and mice possess comparative immunity.

Man.—The character of the local process induced by the diphtheria bacillus in man is not always uniform. "In mucous membranes there may be a simple redness or a catarrhal inflammation; there may be a fibrinous exudate which infiltrates the membrane; or, intermingled with pus cells, epithelial cells, red blood-corpuscles, bacteria, and granular matter, it forms a thick or thin pellicle on the affected surface. This pellicle may undergo coagulation necrosis, and hand in hand with this there may be superficial or deep coagulation necrosis of the mucous membrane."

Virulent diphtheria bacilli remain for a variable length of time in the mouths of persons who have recovered from diphtheria. In the researches of Park and Beebe, in one half the cases the bacillus could not be demonstrated by culture three days after disappearance of the false membrane. In four cases out of six hundred and five examined, it remained in the throat four weeks; in two of the cases, nine weeks.

Some importance attaches to the observation that virulent diphtheria bacilli may be harbored in the throats of persons attending diphtheria patients. Although they themselves show no evidence of the disease, they may act as carriers of infection to susceptible individuals.

Secondary and mixed infection in diphtheria have received considerable attention since the introduction of diphtheria antitoxin. The staphylococcus pyogenes aureus and albus and the streptococcus pyogenes are the most common associated micro-organisms. The diplococcus of pneumonia and Friedlaender's pneumonia bacillus are occasionally met with. The diphtheritic lesion is the channel through which secondary infection of internal organs by the pyogenic cocci occurs.

Toxic Products.—The pathogenic action of the diphtheria bacillus depends upon the absorption of soluble poisonous compounds formed during its growth. These compounds belong to the toxalbumins, and are freely soluble in water. They are not produced by chemical changes in the proteid constituents of the culture medium, but are products of the bacteria themselves. The amount of toxin production differs greatly in cultures from different sources. It is abundant in those cultures in which a secondary alkaline reaction takes place, but is almost nil in those having a permanent acid reaction.

Diphtheria Antitoxin.—The first successful immunization experiments in diphtheria were those of Ferran, published in 1890. To Behring and Roux belongs the credit of the discovery that the blood-serum of an immunized body would protect against the diphtheria bacillus and its poisons. The protective element has been called "antitoxin." It has not been isolated in a pure state, and its chemical nature is unknown.

Method of Preparing the Toxin.—The first step in the development of diphtheria antitoxin is the production of a strong toxin. For this purpose the most virulent bacillus obtainable is grown at 37° C. in 4 per cent alkaline peptone bouillon. In order to obtain the toxin the bouillon is filtered, after a variable length of time, through a Pasteur-Chamberland filter, which rids it of all bacteria.

The strength of the toxin is determined by injections into guinea-pigs, but there is no uniform method of stating the strength. Some take a guinea-pig weighing 500 grams and estimate the toxin strength in amounts required to kill a pig of this weight, while others estimate the amount of toxin required for each 100-gram-weight of pig.

The standard toxin of Behring was of such strength that 0.01 c.cm. to 0.1 c.cm. would produce death in a 500-gram guinea-pig in twenty-four to forty-eight hours. The production of toxins of this strength required five to six weeks' development of the bacillus in 2 per cent peptone bouillon. Cultures of the diphtheria bacillus have since been obtained which give in one week a toxin of 0.005 strength. For the preservation of the toxin carbolic acid, trikresol, or formaldehyde is used. It slowly loses strength with keeping.

Antitoxin is produced by injecting into animals in constantly increasing doses the diphtheria toxin thus prepared. The horse is usually selected, on account of the comparative ease with which immunization can be produced and the large yield of serum. The initial dose of the standard toxin is about 1 c.cm. It is followed by local reaction and by fever and other constitutional symptoms; these usually disappear on the fifth or sixth day. Constantly increasing amounts of the toxin are injected as rapidly as the animal is able to stand them. Ordinarily the injections are made every six or eight days. After about three months as much as 300 c.cm. of virulent toxin can be borne by the horse without any symptoms.

To obtain the serum the animal is bled from the external jugular vein. As a rule, about one quart is withdrawn at each bleeding. The blood is allowed to clot, and is set aside in the refrigerator for twenty-four to forty-eight hours, in order to allow the serum to separate. The clear serum is then pipetted off and preserved by the addition of carbolic acid, trikresol or formaldehyde. Before it is bottled the serum is again filtered through unglazed porcelain. It is now ready for use. Antitoxic serum should be as fresh as possible. It rapidly deteriorates, and after six months its potency is greatly diminished.

The strength of diphtheria antitoxin is usually estimated according to Behring's standard. The normal serum of Behring was of such strength that 0.1 c.cm. would protect a 500-gram guinea-pig against ten times the minimum fatal dose of diphtheria toxin. "Each cubic centimeter of this serum he called an *immunizing unit.*" Further experiments having demonstrated that a much stronger serum could be produced, Behring termed as the *antitoxic unit* one cubic centimeter of an antitoxic serum of such strength that 0.01 c.cm. is sufficient to protect a 500-gram guinea-pig against ten times the fatal dose of diphtheria toxin. The antitoxic unit, therefore, is ten times higher than the immunizing unit. The strength of serum is measured, according to Behring's standard, in antitoxic units. The antitoxic possibilities of serum were not reached by any means in this stronger preparation. Antitoxins have been produced which contained 800 to 1,000 antitoxic units per cubic centimeter. For therapeutic uses they should have a strength of not less than 250 antitoxic units per cubic centimeter, and preference should be given to even more concentrated serum.

Diphtheria antitoxin has little or no effect upon the germ itself. If immunization be produced with the living germ instead of the toxin, antitoxin is formed in very small quantity, but the bactericidal property of the serum is very great.

Attention has been called by Wasserman to the fact that an antitoxic substance is occasionally present in the blood-serum of healthy adults who have never had diphtheria. Bolton and others have observed that some horses are but little susceptible to large doses of the diphtheria toxin.

Antitoxin has been produced from cultures of the diphtheria bacillus by means of the electric current.

An attempt has been made to immunize horses against both the streptococcus pyogenes and diphtheria toxins. Horses immunized against diphtheria are strongly resistant to streptococcus infection. It is also a fact that diphtheria antitoxin acts well in streptococcus diseases.

PSEUDO-DIPHTHERIA BACILLI.

The designation *pseudo-diphtheria bacillus* originated with Loeffler, who applied the term to a bacillus found in diphtheritic false membrane and also in the healthy mouth and throat. Micro-organisms resembling the bacillus of diphtheria are widely distributed. They are not only present in a small per cent of healthy throats, but are met with greater regularity on the normal and diseased conjunctiva, being found almost in pure culture in xerosis. They are also observed in acute and chronic conjunctivitis, trachoma, and ulcer of the cornea. Pseudo-diphtheria bacilli have also been found in ozena, in impetigo and acne pustules. The "flask bacillus" of Unna belongs to this group.

Marked differences are presented by cultures of pseudo-diphtheria bacilli; but these are not greater than the variations seen in pure cultures of the true diphtheria bacillus. There are pseudo-diphtheria bacilli which form upon the surface of agar a thick, creamy coat, and others which show only slight development and are not distinguishable from the true form. Upon potato, growth is similar in both. Morphologically the pseudo-diphtheria bacilli are larger and plumper. They do not change the reaction of the medium.

There is no experimental evidence which would warrant the belief that the bacilli of this group are derivatives of the true diphtheria bacillus, or that they may under fortuitous circumstances become virulent diphtheria bacilli. They may multiply in the body, setting up various processes, but not true diphtheria. They are devoid of toxic action upon lower animals.

BACILLUS OF TETANUS.

Tetanus is a toxic disease, due to the absorption of poisonous products of a specific bacillus. The bacillus remains limited to the point of infection.

The tetanus bacillus, which was discovered in 1884 by Nicolaier, and obtained in pure culture by Kitasato in 1889, is a widely distributed saprophyte. It is found in earth, manure, normal intestine, and in the wounds of animals infected with tetanus.

Morphology.—Slender rods, 0.3 μ to 0.5 μ by 2 μ or 3 μ. It sometimes, by the union of several members, forms long filaments. Motility is present if oxygen be excluded. Spore formation takes place in thirty hours at body temperature. The spores form in one end of the bacillus, giving them a "drum-stick" appearance.

Growth on Culture Media.—As the tetanus bacillus is a strictly anaerobic variety, it requires special methods for cultivation. (See this chapter.)

On gelatin plates it shows a slow growth. The center of the colony is dense and yellow; from it delicate filaments, made up of bacteria, extend in every direction.

Gelatin Puncture Culture.—Short filamentous processes extend in a radiating manner from the puncture canal into the gelatin, which begins to liquefy in two to three weeks. There is slight gas production.

Agar.—Surface growths are characteristic. To the naked eye they appear cloudy. Microscopically the growths are seen to be made up of minute filamentous shreds, intermingling freely. This characteristic shred formation does not take place in any of the other anaerobic species.

Bouillon.—A slight opacity is brought about.

Blood-serum is slowly liquefied by the tetanus bacillus.

Milk.—The growth is luxuriant; it does not cause coagulation.

Irrespective of the medium upon which it grows, the tetanus bacillus gives off gases of a disagreeable, characteristic odor.

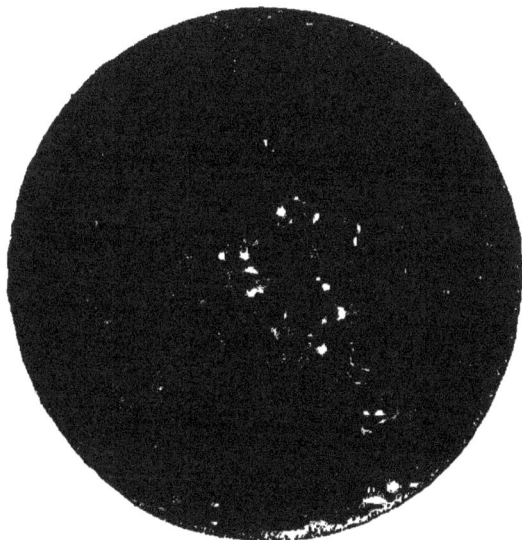

Fig. 23.
Bacillus of tetanus, bearing spores. X 1000. Pfeiffer.

The most rapid growth takes place at body temperature. It ceases at 14° C.

Vitality.—The spore-bearing bacillus is able to resist a high temperature, 80° C., for an hour. Steam destroys it in ten minutes. The spore lives for months in the dry state.

The bacillus also resists weak carbolic acid solutions, and will grow in media of a slightly acid reaction. The presence of oxygen is detrimental to its growth—it is strictly anaerobic. In the presence of the pneumococcus it can be made to grow in oxygen. It has also been made to grow in oxygen by gradually accustoming it to the gas. It is said to lose its virulence under this condition.

Method of Isolation.—From the fact that the bacillus is anaerobic and is usually found in combination with large numbers of other micro-organisms, isolation is difficult.

To destroy the associated bacteria, Kitasato, after spreading the suspected material on agar, placed it in the incubator for twenty-four hours, so that all aerobic species would develop. At the end of this time it was heated to 80° C. for one half to one hour. This destroyed all bacteria and all spores, excepting those of the tetanus bacillus. After this procedure, anaerobic cultures were made in the customary way, by the exclusion of oxygen.

The bacillus can be isolated from the soil, in the upper layers of which it is abundant, by introducing a bit of garden earth or dust under the skin of a mouse or guinea-pig. The animal dies in one to three days. Anaerobic plates are made from around the seat of inoculation, and as a rule the tetanus bacillus can be recovered.

Staining.—The bacillus stains readily with the simple dyes and with Gram's method. The method of staining spores is given in a previous chapter.

Toxic Products.—Brieger isolated from impure cultures four toxins: (1) tetanin, very poisonous, which causes tonic and clonic convulsions; (2) tetanotoxin, similar in action but slower and less virulent; (3) spasmotoxin, identical in action with tetanotoxin; (4) an unnamed basic substance which causes in addition to the muscular symptoms profuse salivation. These substances are less powerful in action than another poison which has since been isolated, and are probably secondary products of the true toxin resulting from the action of the reagents used in separating them.

There has been isolated from bouillon cultures a most powerful substance which, no doubt, is the true poison. It has been obtained in an almost pure state, but its chemical nature is still uncertain. It is very probably not an albuminous body. Brieger and Cohn state that 0.00000005 gram of their purified tetanus toxin is fatal to a fifteen-gram-weight white mouse in four to six days.

Pathogenesis. — *Lower animals* nearly all succumb to small quantities of the germ or its poisons in one to three days with typical symptoms of tetanus.

Mice, guinea-pigs, horses, and rabbits are most susceptible. Pigeons and chickens are slightly so. Amphibia are immune.

Little or no local reaction follows the injection of a small quantity of pure culture subcutaneously, and the bacillus is seldom to be found in the tissue. After a period of incubation of one to three days the muscles near the point of inoculation become the seat of tetanic convulsions. The entire body is gradually involved, and death usually results. The fact that the bacillus is seldom present in the infected tissue is evidence that it does not multiply in the body, and led to the supposition that the pathological action is dependent on a poisonous by-product. Pure cultures, from which the toxic products have been separated, when injected under aseptic precautions have no effect. They die in the body.

It is probable that natural infection is almost universally the result of the introduction of tetanus spores. There are certain conditions necessary in order that they may develop and elaborate the toxin. The wound must be of such a character that oxygen is excluded, and there must be injury to the tissues or concurrent infection by other pathogenic or by non-pathogenic organisms.

The introduction of spores in pure culture into healthy tissue is not followed by tetanus for the reason, it is supposed, that active phagocytosis occurs and they are destroyed. That concurrent infection with other species must occur is shown by the experiment of introducing certain saprophytes with the spores. This enables them to develop and cause tetanus. In experimental tetanus produced by the injection of earth a large number of saprophytes are always present in the exudate. If from the pus of the first case a second inoculation be made, the number of saprophytes is less, and after three or four inoculations very few non-pathogenic species develop, and tetanus, if it occurs at all from the latter injection, is a slight affection.

As before stated, the tetanus bacillus remains at the point of entrance, symptoms arising from the absorption of the toxin. Absorption is very rapid, occurring in forty minutes in the experiments of Roux and Vaillard. That the blood and serous fluids contain the toxin is shown by the development of tetanus after their introduction.

The poison affects the cerebro-spinal system, causing increased reflex excitability of the motor ganglion cells of the spinal cord and medulla. The sensory nerves are not affected. Elimination of the toxin occurs with the saliva and urine.

Tetanus which comes on without an apparent injury is thought often to be due to absorption of toxins through slight abrasions in the alimentary canal, in which the spores are normally present, and under some conditions vegetation and production of toxin may occur. Small skin lesions which escape notice may also explain such cases.

Feeding and inhalation experiments with the tetanus bacillus have never produced the disease.

Immunity Experiments. — Immunity to the toxic products of the bacillus of tetanus has been brought about artificially. The poison, very much attenuated by the addition of trichloride of iodine or Gram's solution, was first injected. The strength was then gradually increased until the body was able to stand large quantities of the bacteria or the toxins.

The serum (antitoxins) of animals artificially immunized acts as a protective if injected simultaneously with or before the toxins. If injected after the body has been infected, the experiment fails.

Tetanus Antitoxin—For the production of the antitoxin the bacillus is cultivated in bouillon in an atmosphere of hydrogen, and at the end of two weeks filtered through porcelain. The filtrate contains the tetanus toxin, and is injected into horses, beginning with about one half cubic centimeter. A horse well immunized against tetanus is able to withstand seven to eight hundred cubic centimeters of the virulent toxin. The strength of tetanus antitoxin is measured in tetanus antitoxic units; that is, the amount of serum necessary to protect one million grams of white mice constitutes one tetanus antitoxic unit. The antitoxin should be of a strength of one to three to five millions; in other words, one cubic centimeter should protect three to five million grams weight of white mice against the fatal dose of tetanus toxin.

Immunization of man has not had a sufficient test to judge of its merit, although its use as a prophylactic measure appears plausible.

For the immunization of animals against tetanus, and also as a cure, veterinarians employ the blood-serum of horses convalescent from tetanus. Blood is drawn ten days after symptoms have subsided, and the serum injected intravenously in half-ounce doses as a curative serum.

YELLOW FEVER.

Researches upon the etiology of yellow fever have been made by several observers. The studies of Sternberg in Brazil, Havana, and other places were conducted according to modern methods. He found in the intestinal tract, and also in the blood of a considerable proportion of yellow-fever cadavers, a facultative anaerobic, non-motile, non-liquefying bacillus—his bacillus X—and suggested that it might be concerned in the etiology of yellow fever. But in his experiments at the time the bacillus X did not give rise to the phenomena of yellow fever as observed in the human subject, and he declined to state positively its etiological relation to the disease. Recent experiments with intravenous injections of the bacillus X have given rise in dogs to the pathologic changes of yellow fever.

In 1895 Sanarelli discovered a bacillus in thirty-eight of fifty-four cases examined, which possesses distinct cultural peculiarities and gives rise in animals to morbid phenomena of such a nature that Sanarelli declares it to be the causative element in yellow fever. He suggests for this micro-organism the name *bacillus icteroides.*

As described by him the bacillus icteroides usually occurs in pairs, and is a small, non-motile, pleomorphous rod, sometimes so short as to be easily mistaken for a coccus. It occurs sparsely distributed in clumps in the capillary blood-vessels of the internal organs; rarely in the intestinal contents.

The bacillus icteroides does not liquefy gelatin, and grows both at room temperature and in the incubator—more rapidly in the latter. It ferments glucose and saccharose actively, lactose slowly.

On gelatin plates, transparent, slightly granular colonies are formed. The granular character becomes more marked as the colony grows older, and the central portion becomes quite opaque.

Upon agar, if grown in the incubator, the colonies are "roundish, gray, a little iridescent, transparent, with a smooth surface and irregular margins." Grown at 20° C. to 30° C., "the colonies are like drops of milk, opaque, prominent, and with pearly reflections." If the culture upon agar is grown from twelve to sixteen hours at 37° C., and then for the same length of time at room temperature, the colonies present an appearance which distinguishes them from all other bacteria. "The colonies then show a flat central nucleus, transparent and bluish, surrounded by a prominent and opaque zone."

Sanarelli's bacillus is able to resist drying for a greater length of time than most other non-spore-bearing pathogenic bacteria. It is killed in seven hours by sunlight. It lives for a long time in sea-water.

Development of the yellow fever bacillus upon culture media seems, according to Sanarelli, to be favored by the presence of certain common mould fungi. He observed that when these moulds developed upon plates which had previously remained sterile, typical colonies of the bacillus grew around them.

Staining.—The bacillus icteroides stains with the simple aniline dyes, but not by Gram's method.

Toxic Products.—Sanarelli has prepared by filtering bouillon cultures a toxin which, injected into the circulation, sets up all the clinical symptoms of yellow fever. The poison withstands 70° C., but is attenuated by boiling.

Pathogenesis.—The bacillus is pathogenic to all the laboratory animals. White mice die five days after injection. The principal change is a fatty degeneration of the liver. The bacillus is found in the blood-vessels. In guinea-pigs it produces "a cyclical febrile disease, which always ends fatally after seven or eight days." Intravenous injection sets up in dogs a process identical in many respects with human yellow fever.

There can at present be very little doubt that invasion of the blood-vessels by the bacillus icteroides is the cause of yellow fever. The mode of entrance is undetermined, but in all probability it is through the respiratory tract and not the intestinal canal. As before stated, the specific microbe is present in the capillary blood-vessels. A toxin is elaborated which is responsible for the constitutional symptoms. It is difficult to isolate the bacillus from the blood during life, or from the cadaver except in those cases which die at the end of the yellow fever cycle—the seventh or eighth day of the disease. In these cases the bacillus invades the blood-vessels in large numbers. Rarely, however, is it found alone; its poisonous products have such an effect upon the intestinal tract that various microbic species are able to invade the blood-vessels even during life, secondary infection by the bacillus coli or the pyogenic cocci being the rule in fatal cases.

HYDROPHOBIA (RABIES).

Hydrophobia is a toxic disease. The origin of the toxic substance is supposed to be in some germ, although a specific bacterium has not been observed in the disease. The larger animals and man are susceptible.

The disease-producing substance, be it a poison or living bacterium, is introduced with the saliva by the bite of an animal, usually a dog, affected with the disease. After the infection the wound heals readily, and a period of ten days to six months intervenes before the outbreak of the symptoms—dread of water, excitement, convulsions, etc.—which characterize the disease. At the outbreak of these symptoms, a reaction, redness, and sometimes suppuration, takes place in the wound.

Diagnosis of Rabies.—The disease can be transferred from one animal to another by inoculation of a small quantity of the brain or spinal cord of an affected animal. In diagnosing the affection subdural inoculation of the rabbit is done with a watery emulsion of the brain or spinal cord of the suspected animal. As a rule the first sign of illness is hyperesthesia and partial paralysis of the hind limbs, which comes on in a period varying from fifteen to thirty days. The incubation stage, however, may be of shorter duration, or it may be delayed as long as three months. The paralytic stage lasts a variable length of time—sometimes a few hours, sometimes a few days—gradually increasing until the animal is unable to move and exhibits no other sign of life than infrequent shallow respiration. A slight rise of temperature usually precedes the appearance of paralytic symptoms, to fall again to normal in forty-eight to seventy-two hours. Rigor mortis is very pronounced. At the post-mortem examination the trephine wound is found to be healed, the brain and internal organs normal. The tissues contain no bacteria.

Immunity Experiments.—Hydrophobia is of interest to the bacteriologist on account of the successful experiments at immunization which have been made by Pasteur.

He found that by drying the nerve substance (in dark place) the disease-producing property was much diminished. The dried cord is triturated with bouillon and used to immunize animals. The first injection is small. The amount is gradually increased until the animal is able to withstand large doses of the original or unattenuated emulsion.

The method is applied to human beings bitten by rabid animals. Owing to the long stage of incubation in hydrophobia, it has been used with success after infection has taken place. Passive immunity by the injection of serum of immune animals has not been successful.

BACILLUS OF BUBONIC PLAGUE.

Bubonic plague is an infectious disease, occurring principally in the Orient, caused by a specific germ and characterized by suppurative inflammation of lymphatic glands throughout the body.

The bacillus of bubonic plague (bacillus pestis) was discovered at about the same time, in 1894, by Kitasato and Yersin.

Morphology.—A short, thick bacillus with rounded ends, forming in cultures long chains. In the tissues it is surrounded by a delicate capsule.

The bacillus does not form spores.

Growth on Culture Media.—Upon *blood-serum*, at incubator temperature, there occurs an abundant growth of a yellowish-gray color. The medium is not liquefied.

Colonies upon *glycerin-agar* are of a whitish-gray color, and by reflected light have a bluish appearance; under the microscope they appear "as if piled up with glass wool; later as if having dense, large centers."

Development in *bouillon* causes cloudiness, with a deposit of bacilli upon the sides and bottom of the test-tube—an appearance resembling the growth of the streptococcus pyogenes.

A slight growth occurs upon *potato* at 34° C., but not at room temperature. It is thin, dry, and of a whitish-gray color. Growth takes place most luxuriantly at 38° or 39° C.

Vitality.—Direct exposure to sunlight destroys the bacillus in three hours.

Drying one to four days, at room temperature, is sufficient to destroy it. Bouillon cultures are sterilized in thirty minutes by exposure to 80° C. ; at 100° C. they are killed in a few minutes.

Method of Isolation.—In bubonic septicemia the bacillus can be readily isolated from the blood by pricking the finger and allowing several drops to flow into an agar tube. Colonies develop in twenty-four hours.

The bacillus can also be obtained from the diseased lymphatic glands, in which it is present in large numbers.

Staining is easily accomplished with the simple aniline colors, but not by Gram's method. The ends of the rod stain more deeply than the central portion, leaving a lighter space in the middle.

Pathogenesis. — Mice, rats, guinea-pigs, and rabbits are susceptible both to inoculation and feeding. During the prevalence of bubonic plague, mice and rats contract the disease spontaneously. After subcutaneous inoculation death takes place in from two to five days. The post-mortem reveals a reddish, gelatinous exudate at the site of inoculation; the spleen is enlarged; bacilli are found in all the organs. Pigeons are immune. Virulence is rapidly increased by passing the bacillus through guinea-pigs.

In man the plague bacillus enters the body through lesions, often inappreciable, in the skin. The glands in the neighborhood begin to enlarge, and, if the infection is not too virulent, the process may cease at this point; as a rule, however, bacilli find their way into the blood current, setting up septicemia. The bacillus is very abundant in the urine and feces.

In still other cases infection takes place through the lung. In these pneumonia develops and the plague bacillus appears in the sputum.

Infection through the intestinal canal is believed not to occur.

Immunity Experiments.—The blood-serum of persons convalescent from the plague, and also of animals inoculated with progressively increasing doses of the living germ, contains a substance which has a specific influence upon the bacillus.

Yersin and others have produced a curative serum. The method consists in the injection of living bouillon cultures into horses. Yersin produced a serum which, in doses of one and a half cubic centimeters, was sufficient to cure mice twelve hours after infection. Warm cultures are better adapted to the production of the serum than cold ones. Experiments with this serum in the human subject seem to show that it is both protective and curative.

BACILLUS OF MALIGNANT EDEMA.

The edema bacillus (bacillus edematis maligni; vibrion septique), discovered by Pasteur in 1878, is a frequent inhabitant of the upper layers of the soil, particularly in fertilized earth. It is also present in dust and drain-water, and frequently in the intestines of animals.

Morphology.—A rod with rounded ends, as long as the anthrax bacillus, 1 μ to 3 μ—exceptionally 5 μ—and 1 μ to 1.2 μ thick. It is consequently more slender than the anthrax bacillus with which it has been confounded.

The bacillus is slightly motile and has a number of flagella on the sides and poles.

It is usually found in pairs, pole to pole, but also occurs in chains of more members.

Spores develop in the middle of the cell (clostridium).

Growth on Culture Media.— The bacillus is strictly anaerobic, and must be cultivated by special methods previously described. It grows on the ordinary media.

Upon *gelatin plates* the colonies appear as white, irregular bodies, which under a low magnifying power are irregularly marked with radiating and interlacing lines.

The bacillus grows best at incubator temperature.

Method of Isolation.—A susceptible animal is inoculated with garden earth containing the bacillus, and when the disease has developed a small quantity of the effusion in the areolar tissue is stirred into liquefied gelatin and the medium allowed to solidify. Numerous colonies develop in the depth where the oxygen does not reach. They liquefy the gelatin, rapidly forming small spherical cavities in which the bacilli are mixed with the fluid gelatin. The fluid appears slightly cloudy. Some of the contents of one of these cavities can be transferred to gelatin plate or tube, and pure culture obtained by growing in hydrogen.

According to Kitasato the bacillus can be isolated without fail from the spleen of a guinea-pig which has succumbed to the disease, by bringing a small bit into bouillon made of guinea-pigs and growing in hydrogen.

Staining.—The bacillus stains with the simple dyes, but is decolorized with Gram's method.

Pathogenesis.—*Lower Animals.*—Nearly all of the lower animals are susceptible to inoculation; mice especially so. The ox is said to be immune.

Ordinary subcutaneous injections do not cause infection, as the oxygen present interferes with development. Infection in susceptible animals is successful if the subcutaneous tissue is broken up with a pair of forceps through an opening of the skin. Fertilized earth, in not too small a quantity, or bits of pure culture are introduced deep into the pocket. Death results from this procedure in one or two days.

At the autopsy a widespread subcutaneous edema is found, the exudation containing great numbers of the bacilli. They seldom develop in the internal organs or the blood during life, owing to the oxygen present. Infection takes place so rapidly in the mouse that in this animal the bacillus frequently multiplies in the blood-vessels. After death it also enters the circulatory system in other animals.

Man.—The disease has been observed in man, but only in individuals who have from fever or other causes become debilitated.

Immunity Experiments.—Several species of animals (horse, sheep, dog) have been successfully immunized by injection of toxic bacterial products. (Chamberland and Roux.)

BACILLUS OF SYMPTOMATIC ANTHRAX.

Symptomatic anthrax (black leg) is an infectious disease occurring principally in young cattle. It occurs during the months that the animals are at pasture. It has been found in all parts of the world as a sporadic disease, occurring also as an epidemic in certain regions. The disease is characterized by swelling of the skin and muscular tissues, particularly of the legs and quarters, hence the name "quarter evil." It shows great similarity to localized anthrax, and they were formerly considered one and the same disease.

Bacillus of Symptomatic Anthrax (bacillus des Rauschbrand; bacillus sarcophysematos bovis), the specific germ of the disease, is always present in the affected tissue. It was discovered by Feser and Bollinger, and obtained in pure culture by Kitasato.

Morphology.—A long, slender rod, 3μ to 6μ by 0.5μ to 0.7μ, with rounded ends. The cell usually occurs singly in culture.

It is supplied with numerous flagella, and has limited motility. In artificial culture there is a tendency to degenerate and produce involution forms.

It forms oval spores which are thicker than the mother cell. The spores hold a position in the mother cell a little to one side of the middle.

Growth on Culture Media.—It is strictly anaerobic. Growth takes place most rapidly at 37° C. It ceases below 14° C.

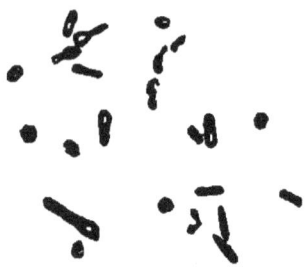

Fig. 24.

Rauschbrand bacillus.

Cultures in *gelatin*, made according to the method of Liborius, liquefy the medium and cause the formation of globular cavities, from which threads of bacilli sprout into the gelatin.

When grown on *agar* or *blood-serum* a peculiar rancid odor is given off.

It grows well in slightly acid bouillon.

The growth on *potato* (in hydrogen) is similar to that of the typhoid bacillus. A transparent, colorless film covers the surface.

Vitality.—The bacillus retains its vitality on solid media indefinitely.

It is destroyed in five minutes when exposed to 100° C. The bacillus shows considerable resistance to chemicals, being able to live in five per cent carbolic-acid solution for ten hours.

Method of Isolation.—Pure cultures are made from the blood and organs of infected animals by the usual methods of cultivating anaerobic species.

Staining.—Stains with the simple dyes. Gram's method decolorizes it.

Pathogenesis.—Most *lower animals* are susceptible to the action of the bacillus, especially cattle, goats, sheep, and guinea-pigs. Pigs, dogs, cats, rabbits, rats, and chickens are almost immune. Frogs can be infected by elevating their body temperature to 22° C.

Infection usually takes place through wounds. Death follows the infection in twenty-four to thirty-six hours. The tissue (as seen at autopsy) is infiltrated with a sanguineous fluid of a reddish-black appearance, containing many bacilli. The diseased area is emphysematous, owing to gas produced during growth of the bacillus; hence the German name for the affection (*rauschen*, to crackle). The bacillus multiplies in the body after death, and is found throughout the tissue.

It has been shown that the addition of a twenty-per-cent solution of lactic acid to the culture increases its virulence.

Man is not affected.

Immunity Experiments.—According to Kitt the bacteria and their poisons are attenuated by exposing the flesh of an animal which has died of the disease to 100° C. for six hours in a steam sterilizer. By the injection of this attenuated virus he immunized sheep, guinea-pigs, and oxen.

Protective inoculation, after the method of Pasteur, is applied, and has been found serviceable for the immunization of cattle and sheep.

An immunizing serum is on the market. It is efficient but expensive.

BACTERIA WHICH PRODUCE SEPTICEMIA.

ANTHRAX BACILLUS.

Anthrax (splenic fever; charbon) is probably the oldest disease of domestic animals of which we have any record. It is supposed to have been the murrain of cattle recorded in sacred history; it was a common and fatal disease among the oxen at the siege of Troy. Virgil, whose poems display considerable veterinary knowledge, has left us a clear account of the disease as it prevailed in his time in Italy. At the present day anthrax is a common disease in France, England, India, and Austro-Hungary. In Siberia and parts of Russia, on account of its extensive prevalence, the disease is known as the Siberian Pest. In the United States anthrax is most common in the southern part of the Mississippi Valley, and is very fatal to mules and horses as well as to cattle. It seems to be on the increase with us.

The Bacillus of Anthrax.—The cause of anthrax is found in a specific germ, the bacillus anthracis. Considerable historic interest attaches to this micro-organism. It was discovered by Pollender in 1849. Davaine, whose researches were published in 1863, first appreciated the true significance of the presence of the bacillus in tissues. Koch cultivated it upon artificial media, and transmitted the disease to susceptible animals.

Apart from the diseased tissue and the discharges of infected animals the anthrax bacillus is rarely found, although it is capable of leading a saprophytic existence in water and in the soil.

Morphology.—A large, non-motile bacillus, varying in width from 1 μ to 1.25 μ, and in length from 5 μ to 20 μ; occurring singly in the blood of animals, but growing out into long filaments in culture. The bacillus upon different media varies in size. In cultures the ends are more or less rounded, but in the animal body they appear concave, leaving a lenticular interspace between them. It is a spore-forming organism, and on account of the large size the cell is suitable for the study of this process in the hanging drop.

Growth on Culture Media.—*Upon gelatin plates,* at 24° C., colonies appear at the end of twenty - four hours. They are small, white, and opaque. Upon the surface growth occurs with moderate rapidity, and the colonies have a yellowish-white color. "The central portion is surrounded by a mass of wavy filaments which are associated in tangled bundles." Liquefaction of the medium begins on the third day.

Fig. 25. Anthrax bacillus.
Contact preparation.
× 1000. Fraenkel and Pfeiffer.

In gelatin puncture cultures growth is most abundant upon the surface. In the depths of the medium it is composed of numerous filaments which extend out into the culture medium. Liquefaction begins on the third or fourth day. It is noticeable first at the surface, and gradually extends downward, the whitish mass of bacilli sinking in the liquefied gelatin, leaving the upper layer clear.

Colonies upon agar plates present an appearance similar to those on gelatin. They are visible after twenty-four hours in the incubator. The growth in stab cultures is also similar to that in gelatin. Upon stroke cultures the growth is grayish-white and shining.

On blood-serum it is white and not characteristic. This medium is liquefied.

Development is abundant upon *potato.* The mass of bacilli has a yellowish-white color, and is very rich in spores.

Bouillon is clouded; the bacilli finally sink to the bottom of the tube, and the medium becomes clear.

Vitality.— The non-spore-bearing anthrax bacillus is destroyed in ten minutes by a temperature of 54° C.; the spore has remarkable resisting powers to heat and chemical agents. In the absence of moisture it is destroyed by

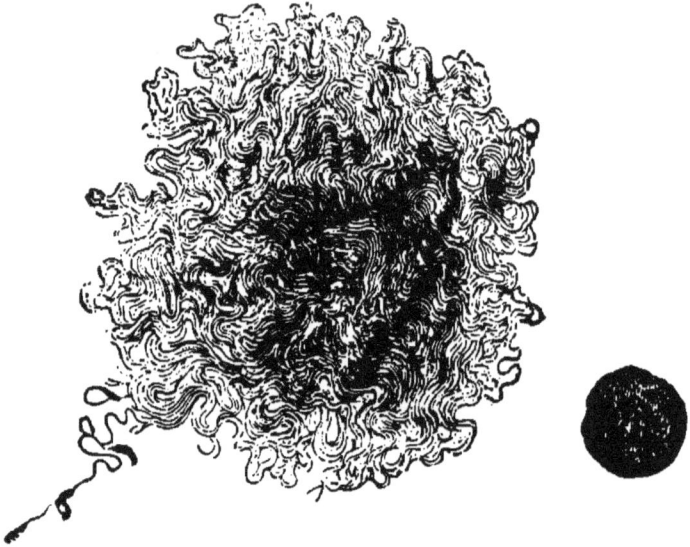

Fig. 26. Colony of anthrax bacillus. × 80. Flügge.

exposure for three hours to 140° C., "but spores suspended in a liquid are destroyed in four minutes by 100° C."

Spores can be preserved almost indefinitely in the dry state without losing the power of germinating when brought again under suitable conditions.

Spore formation takes place in artificial media when free access of oxygen is permitted, as on the surface of solid culture media. It does not occur in the animal body during life. A temperature of 20° to 35° C. is required, 30° C. being most favorable. Non-spore-forming varieties have been produced by cultivation in carbolic acid in sufficient quantity to lessen the vigor of development (1–1,000).

Staining can be easily accomplished with the simple watery solutions, and also by Gram's method. In the latter process the action of the iodine upon the cell protoplasm causes it to lose its homogeneous structure, and the coloring matter often appears in clumps with unstained intervals between them.

Pathogenesis.—Like all pathogenic bacteria, the bacillus of anthrax possesses a varying degree of virulence. Its virulence can be diminished or increased by appropriate means. The disease-producing power of the vegetative germ is transmitted to the spore. The latter may be kept in a dry state for years without undergoing any deterioration as regards its pathogenic power.

A fixed degree of virulence possessed by the bacillus can be maintained by transplantation at frequent intervals and keeping at room temperature.

Cultivation in the body of naturally immune animals destroys virulence: thus, by inoculating a frog and maintaining it at a temperature sufficient to enable the bacillus to develop, the animal is killed. The bacillus, however, loses its pathogenic power for other animals.

Another method of attenuating and fixing virulence is by cultivation at a temperature sufficiently high to retard full development. At 42° C. the bacillus shows marked attenuation at the end of three weeks. Higher temperatures produce the same effect in a shorter time.

The anthrax bacillus increases greatly in virulence with each passage through a susceptible animal. By rapid inoculation Watson-Cheyne obtained a culture of such strength that a single germ (estimated) was sufficient to induce a fatal 'infection in the guinea-pig.

The anthrax bacillus is pathogenic to all herbivorous animals; carnivora as a rule are exempt. Man belongs to the slightly susceptible animals. Among the rodents used for experiment the white mouse is most acutely susceptible, the white rat but slightly so. The guinea-pig, which is highly susceptible to subcutaneous inoculation, is refractory to intestinal infection with the fully developed bacillus or spore. Among the domestic animals, horses, cattle, sheep, and goats are highly susceptible to the disease.

Infection in the natural way usually takes place through the alimentary canal. The vegetative cell is killed by the acid stomach juices, but the spore is not affected. Passing through the duodenum unharmed, in the alkaline fluids of the small intestine it finds conditions favorable to rapid vegetation. The bacillus penetrates the mucous membrane of the intestine, and, pushing aside the epithelium of the villi, finds its way into the blood-vessels of the gut. From these it is distributed throughout the body. In the circulating blood, during life, the bacillus occurs very sparingly, but if the body be allowed to remain unopened a few hours after death, very rapid multiplication of the bacillus takes place.

The gross pathological changes are often very slight. Beyond enlargement of the spleen, and sometimes hemorrhagic spots on the peritoneum and pleura, the internal organs appear to be normal. If, however, a piece of the spleen, liver, kidney, or lung is hardened in alcohol and stained by Gram's method, the bacillus will be found in enormous numbers in the capillary blood-vessels — a true septicemia is present. Inhalation of anthrax spores is followed in cattle, and also in the human subject, by fatal pneumonia. As a rule pulmonary infection does not lead to septicemia. In the human subject this form of pneumonia was, before the discovery of its true nature, often called "wool-sorter's disease," on account of its occurrence among persons engaged in handling the wool and the skins of sheep which had died of anthrax. Local infection of the skin in the human subject is called "malignant pustule." It is sometimes followed by septicemia, and is fatal in about fifty per cent of all cases.

Toxic Products.—Considerable research has been made into the nature of the toxic substances produced by the anthrax bacillus, both in the animal body and in artificial culture. Notwithstanding the enormous number of bacilli in the blood-vessels after death, it is probable that their presence alone and the appropriation by them of nutritive material can not account for the great fatality from this

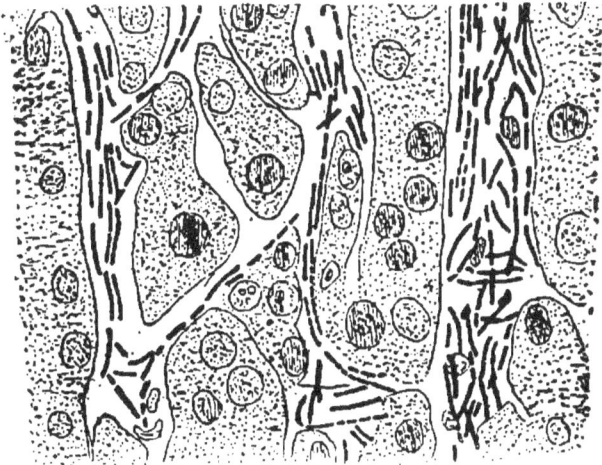

Fig. 27. Anthrax bacillus in tissue. X 700. Flügge.

disease. It is due, in a large measure, to the production of toxins or toxalbumins.

Martin, in his experiments, found in filtered cultures proto-albumose and deutero-albumose, which in small doses in mice caused edema, coma, and death within twenty-four hours. He also isolated an alkaloidal substance which, in doses of 0.01 to 0.15 gram, produced death in two or three hours. Hankin isolated an albumose which was very toxic and capable of conferring immunity in rabbits against anthrax infection.

Immunity Experiments.—In the production of artificial immunity, according to the original method of Pasteur, two inoculations were used. The first or weaker vaccine consisted of anthrax bacilli whose virulence had been diminished by heat. A rise of temperature, and sometimes local edema, followed. Ten or twelve days later he introduced a second vaccine which consisted of a bacillus of such virulence that if used for the primary inoculation it would prove fatal in many cases.

This method granted a partial immunity against anthrax. It protected against local infection, but not against infection by the intestinal canal. It is more effective in cattle and horses than in sheep and goats.

Such inoculations are attended with some danger. As a living germ is introduced, a fatal infection sometimes follows, and in that way the method served for the further propagation of the disease. The "anthrax vaccine," which is at present sent out under the label of the Institut Pasteur to all parts of the world, consists of bouillon cultures of the anthrax bacillus. The method of preparation is kept secret.

The most feasible method of protective inoculation consists in the use of highly virulent cultures which have been sterilized by heat. This obviates the danger of perpetuating the disease, and would in all probability prove more effective, especially in the sheep and goat, than the method of Pasteur.

BACILLUS OF HOG CHOLERA.

Synonyms: Bacillus suis; bacillus of swine plague—Billings; bacillus of English swine fever—Klein.

Klein, of England, described this bacillus in 1884. It is probably one of several different micro-organisms seen by Dettmers in his investigation on the infectious diseases of swine (1879), and called by him bacillus suis. Salmon and Smith, in 1886, isolated it, described fully its biological characters, and established its etiological relation to hog cholera.

Morphology.—The bacillus in liquid media is from 1.2 μ to 1.5 μ in length, and from 0.6 μ to 0.7 μ in breadth. Upon the surface of solid media the cells are almost spherical.

The bacillus is usually single, but chains of three or four members are met with.

It is provided with numerous flagella which can be demonstrated by Loeffler's method, and is highly motile.

Spore formation does not take place.

Growth on Culture Media. — Colonies appear upon *gelatin plates* at the end of twenty-four hours. They rarely exceed one millimeter in diameter, are of a whitish color, and without any distinctive characters.

Stab cultures in this medium grow along the needle track as white, round bodies, which may become confluent. Upon the surface a yellowish-white, abundant layer develops. The medium is not liquefied.

Upon the surface of *agar plates* colonies are much larger, attaining a diameter of four millimeters.

Agar tube cultures are covered over in twenty-four hours with a whitish coat.

Bouillon is uniformly clouded in from eighteen to twenty-four hours.

The growth upon *potato* is abundant and of a straw-yellow color.

When grown in solutions containing glucose, fermentation takes place.

Phenol and indol are not produced.

The hog cholera bacillus grows both at room temperature and in the incubator. It is able to thrive in media having an acid reaction, which it changes to alkaline.

Vitality.—Although the bacillus does not form spores, it is capable of living for a long time, cultures remaining alive for almost a year. It may remain alive for about the same length of time in the soil, but here, after about six weeks, its virulence is gradually lost.

It resists desiccation for a period varying from nine days to several months, depending upon the thickness of the layer exposed for experiment.

Staining. —The hog cholera bacillus stains with the simple aniline solutions, but not by Gram's method. If a very dilute, watery solution of methylene blue be used, the ends of the rod stain while the central portion remains uncolored. For demonstrating the presence of the bacillus in tissue, Loeffler's alkaline methylene blue is suitable.

Pathogenesis.—This bacillus induces in hogs the very common and fatal disease called cholera. Infection takes place through the alimentary canal. In the most acute forms the disease is characterized by more or less enlargement of the spleen and the presence of numerous hemorrhagic spots in the skin, internal organs, and upon the pleura and parietal peritoneum. The bacillus is found in the blood-vessels. In less acute cases the principal changes are in the alimentary canal and the lymphatics of the mesentery. The colon and cecum contain numerous necrotic areas. Sometimes there is a diphtheritic exudate upon the surface. The lymphatic glands are enlarged and discolored. Dropsical accumulations in the peritoneal cavity and broncho-pneumonia are frequent. The bacilli are scanty in the blood, more abundant in the spleen, where they are found in small, irregular groups—the same arrangement that is seen in the spleen of typhoid fever patients.

The rabbit is very susceptible to subcutaneous inoculation : 0.08 c.cm. of a pure culture is usually sufficient to produce death in a week or ten days. The bacilli are found in all the organs, but the principal changes are in the liver, and consist of numerous small, yellowish-white necrotic areas. There is more or less hyperemia of the intestinal mucous membrane, and oftentimes enlargement of the spleen. The guinea-pig dies of septicemia in forty-eight hours after the subcutaneous inoculation of 0.1 c.cm. Pigs are killed by intravenous injection of 1 to 2 c.cm. of living virulent cultures, but usually recover from the same quantity injected subcutaneously. Pigs and fat hogs are most susceptible.

Novy isolated, from old bouillon cultures, a poisonous ptomaine which was fatal to rats in doses of 0.12 to 0.25 c.cm. He also obtained a soluble toxalbumin which was fatal to rats in doses of 0.1 to 0.5 grams.

Immunity Experiments. — In our experiments made with pigs, the only successful method of producing active immunity was found to be the intravenous injection of sterilized bouillon cultures of a highly virulent bacillus. By this method we have succeeded in granting immunity in eighty per cent of hogs exposed (one hundred in number) to a moderately virulent form of the disease.

A certain amount of immunity can be produced in pigs by subcutaneous injection of sterile cultures in gradually increasing quantities, five to six inoculations at intervals of four or five days being required. Pfeiffer pointed out that there is an antitoxin in the blood of animals immunized against the hog-cholera bacillus. He succeeded in producing in the guinea-pig a serum protective in the proportion of one part to three hundred and fifty thousand.

BACILLUS OF CHICKEN CHOLERA.

The bacillus of chicken cholera (bacillus of septicemia hemorrhagica; bacillus of swine plague, of Salmon and Smith; bacillus cholerœ gallinarum—Flügge; bacterium avicidum—Kitt; bacillus der Schweineseuche) occurs in cholera of chickens and geese.

Fig. 23.
Bacillus of chicken cholera.
x 1000.

The disease is characterized by diarrhea. Death results in one or two days. The intestinal contents as well as the blood and viscera contain the specific germ. Its etiological relation to the disease was established by Pasteur, who in 1880 obtained it in pure culture and reproduced the disease by inoculation.

Morphology. — A short, plump, non-motile rod with rounded ends. It is usually found singly, but several may be united.

Spore formation does not take place.

Growth on Culture Media.—It grows on the ordinary media at room temperature.

In gelatin puncture culture a delicate growth extends along the needle track. A delicate grayish-white film also forms on the surface. The gelatin remains unchanged.

On blood-serum and agar a moist white growth.

On potato a very scanty growth, and that only at body temperature.

Staining.—It stains with the simple dyes, but takes the stain only at its poles. It does not stain with Gram's method.

Pathogenesis.—The animals most susceptible to the action of the bacillus are fowls, geese, sparrows, mice, and rabbits. Guinea-pigs are almost immune.

Infection usually takes place through the mouth in chickens, the source of infection being the dejections of diseased fowls. Infection follows much smaller doses of the pure culture given subcutaneously or by intravenous injections than by the mouth.

The principal action of the germ is the production of septicemia, but hemorrhagic enteritis is not uncommon.

The bacillus injected subcutaneously or fed to rabbits causes fatal septicemia characterized by capillary hemorrhages throughout the body, hence the name "bacillus septicemia hemorrhagica" which is sometimes used. Passage through rabbits increases its virulence rapidly.

In Schweineseuche (German) and American swine-plague the lesions consist of numerous necrotic areas in the lungs and fibrinous inflammation of the pleura and peritoneum. Infection probably occurs through the respiratory tract. The bacillus in these diseases presents slight differences in biological characters, but is generally looked upon as identical with the bacillus of chicken cholera.

Immunity Experiments. — When pure cultures are allowed to stand for some time they become attenuated. By the injection of such cultures, followed in a few days by one of a less attenuated culture, and finally by one of highly virulent growth, Pasteur was able to immunize chickens against the disease.

Theobald Smith immunized rabbits by the injection of sterilized cultures.

Metschnikoff brought about the same result in swine with sterile cultures.

BACILLUS OF SWINE ERYSIPELAS.

Synonyms : Bacillus des Schweine-rothlauf; bacillus of mouse septicemia; bacillus of rouget; bacillus murisepticus.

Morphology.—Swine erysipelas bacilli from the organs of diseased animals and from solid and liquid media vary in length from $1.5\,\mu$ to $2\,\mu$, and in breadth from $0.3\,\mu$ to $0.4\,\mu$. In older cultures, notably in bouillon, single filaments are seen measuring $8\,\mu$ in length. The bacillus is non-motile, and does not bear spores.

Growth in Culture Media.— *Upon gelatin plates* colonies are visible to the naked eye on the third or fourth day, and attain full development on the fifth or sixth day. Surface colonies, at the end of this time, have a diameter of three millimeters, and appear as indistinct spots lying in a shallow depression of liquefied gelatin. The deep colonies are a little larger and have not liquefied the medium. Under a magnification of 60 or 80 diameters they are seen to have a central nucleus from which interlacing lines of bacilli radiate in every direction.

In gelatin stab cultures the growth is characteristic and beautiful. "From the track of the wire a delicate growth radiates laterally in all directions. The growth extends slowly until, in two weeks, it has reached the sides of the tube. The medium has then a hazy appearance. Liquefaction takes place slowly down the needle track, and the evaporation following produces a funnel-shaped depression in the medium."

Alkaline gelatin is the most favorable medium for the growth of this bacillus.

Upon agar and blood-serum growth is very feeble, consisting of numerous minute, translucent, discrete colonies.

In alkaline peptone bouillon the bacillus causes a slight cloudiness. When the medium is shaken it takes on a peculiar smoky appearance. Growth is very feeble or entirely absent upon ordinary bouillon. If lactose be added, multiplication is moderately vigorous and the medium takes on a permanent acid reaction.

It does not grow upon *potato*.

The bacillus does not form spores.

Staining. —The swine erysipelas bacillus stains readily by Gram's method. By this method the bacillus is found to be very numerous in the blood-vessels, particularly in the capillaries. Numerous phagocytes are seen loaded with bacilli.

Pathogenesis. — House-mice and pigeons are most susceptible. Swine are probably infected by the alimentary canal. Rabbits are slightly susceptible; field-mice, guinea-pigs, and chickens immune. The mouse septicemia bacillus has been isolated in several outbreaks of disease among swine in this country. In Europe it is a very common and fatal affection. In all susceptible animals this bacillus induces septicemia. The lesions in swine are found principally in the serous membranes, and consist of fibrinous inflammation.

The bacillus is non-pathogenic to man.

Immunity Experiments. —Pasteur experimented upon the protective inoculation of swine against rouget. His method consisted of two inoculations of living cultures, the first being a bacillus weakened by passage through rabbits, and second a bacillus of greater pathogenic activity. The method confers immunity.

BACILLUS TYPHI MURIUM.

Bacillus typhi murium is the name given by Loeffler to a bacillus found by him in 1889, in an epidemic among mice caged for experimental purposes. Death took place in 69 per cent of cases. Infection occurred by the living mice gnawing at those which had succumbed to the disease.

Morphology.—A short, actively motile rod, with numerous flagella on the sides and poles. It sometimes grows in filaments. Spores have never been demonstrated.

Growth on Culture Media.—The bacillus is aerobic and facultative anaerobic. It grows on the ordinary media at room and body temperature.

Gelatin Plates.—Thin, transparent colonies, very similar to typhoid growth. They may be thicker. The medium remains clear and is not liquefied.

Agar.—Grayish-white colonies.

Potato.—White pellicle forms, and the potato around the growth takes on a dirty grayish-blue color.

In peptone bouillon, to which sugar has been added, gases are developed. The medium becomes cloudy and takes on an acid reaction.

Milk becomes acid, but does not change its appearance.

Pathogenesis.—House-mice, field-mice, and white mice are the only animals susceptible to natural infection by the alimentary canal. Death results in one to two weeks.

Nearly all lower animals can be infected by subcutaneous or intravenous injections. Autopsies show enlargement of spleen and hemorrhagic infiltration in the bowel and stomach. The blood and most viscera contain the specific germ. Sometimes it is found in clumps, as in typhoid fever.

As the bacillus typhi murium affects mice but not other animals or man, Loeffler proposed its use for the extermination of field-mice. Bread was saturated with a bouillon culture and scattered about infested fields.

SPIRILLUM OF RELAPSING FEVER.

Relapsing fever (typhus recurrens) has prevailed as an epidemic in Europe, India, and the United States. It is characterized by recurring paroxysms of fever, with complete abatement between the relapses. The constant presence of a very motile spirillum in the blood of those suffering from the disease leaves little doubt as to its etiological relation.

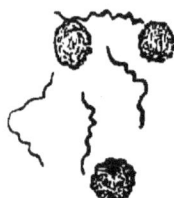

Fig. 29.

Spirillum of relapsing fever.

The spirillum of relapsing fever (spirochaete Obermeieri) was discovered in the blood of individuals during the relapses of recurrent fever. Its presence at this stage was constant, while during the period between the relapses it disappeared entirely.

Morphology.—A slender, spiral cell with tapering ends, whose length varies from 16 μ to 40 μ. Its thickness is 0.1 μ. It has very active spiral motility, and is said by Koch to contain flagella.

Artificial culture of the spirillum has never been successful.

Vitality.—Motility has been retained in blood of leeches for fourteen days (Heydenreich) at 16° to 20° C.

Although spores have not been demonstrated, the entire disappearance of the cells during the intervals of fever would justify the belief that spores are given off during a paroxysm which develop into full grown spirilla by the time the next relapse occurs.

Staining.—The spirillum stains with the simple dyes, but does not retain color when treated by Gram's method. Loeffler's methylene-blue solution is applicable; also the method suggested by Guenther. (See examination of blood.)

Pathogenesis.— Monkeys, by the inoculation of blood of a patient with relapsing fever, have a typical attack after a stage of incubation of three or four days. The monkey has but a single paroxysm.

Man has been inoculated in a similar manner, the fever showing typical relapses.

In fatal cases the spleen is very much enlarged and may contain necrotic areas. The spirilla are not found in the blood after death.

Immunity Experiments have given negative results. One attack of the disease does not protect against another.

The affection known as actinomycosis is found principally in animals of the bovine species; less commonly it is met with in the horse, pig, and the human subject. It is characterized by the formation of hard tumors of variable size, which contain a peculiar fungus — actinomyces, or ray fungus — which is the cause of the disease. The actinomycetic nodules are made up of fibrous tissue more or less dense, in which sooner or later suppuration takes place. The fungus occurs in the pus or imbedded in interstices of

Fig. 30. Actinomyces.

the tissue, and appears to the naked eye as a minute grain about 0.5 millimeter in diameter, varying in color from a pale yellow to a sulphur yellow.

In 1877, Bollinger, of Munich, pointed out the constant presence and etiological relation of the actinomyces or ray-fungus. Israel, the following year, cultivated it from a case of the disease occurring in man.

The *ray fungus* is a streptothrix. Its general structure can be distinguished without difficulty by examination, in the fresh state, of the small, roundish masses which are seen in actinomycotic pus and in the nodules. One of the minute, yellowish grains should be crushed upon a slide, a drop of dilute caustic soda or potash added, and examined with a magnification of 250 to 450 diameters. The fungus is found to be made up, externally, of large, club-shaped bodies which radiate from the central portion of the mass. The latter is made up of numerous very delicate filaments surrounding a central granular substance.

Growth in Culture Media.—The ray fungus is facultative anaerobic, and grows best at high temperatures (35° to 38° C.).

It forms upon *glycerin-agar* round, pin-head-sized masses of a yellowish-brown color and firm consistence; the isolated colonies after a time coalesce, forming large, wrinkled masses.

In bouillon development is abundant and consists of globular masses, always remaining clear.

Upon potato colonies are "dense, yellow and white, circled with black." The potato turns brown.

Blood-serum colonies are similar to those upon agar. The medium is slowly liquefied.

Vitality.—The fungus remains alive and virulent for a year or more upon artificial media.

Staining in tissue can be accomplished by Gram's method.

Pathogenesis.—Actinomycotic lesions may be produced by injection of pure cultures into the abdominal cavities of rabbits and guinea-pigs. Spontaneous infection is seen most commonly in cattle. The ray fungus exerts a varying degree of positive chemotaxis upon the leucocytes, and its multiplication is accompanied by more or less massive inflammatory new growths, which sooner or later undergo purulent degeneration.

Actinomycosis in cattle commonly invades the bones of the jaw, leading to destruction of the osseous tissue. Infection probably occurs in many of these through carious teeth. Curiously enough, invasion of the bones of the face, so common in cattle, is very rare in the human subject. In swine the initial actinomycotic focus is most frequently in the pharynx, on or about the faucial pillars. Actinomycosis of the lung may be confounded with tuberculosis; they have been known to occur as concurrent infections.

The life history of the fungus outside the body is somewhat obscure. It has been found growing upon barley and other plants.

It is believed that actinomycosis is inoculable rather than infectious; that the fungus or its spore is able to live for an indefinite length of time upon plants, and from these it finds its way into a wound of the skin or mucous membrane, or is inoculated into healthy tissues by sharp parts of plants containing it.

MALARIAL FEVER.

In 1880 Laveran presented the results of the study of the
blood in forty-four cases of intermittent and remittent fevers,
and described as occurring in the blood of all these cases
peculiar bodies which he believed to be parasites and the
cause of the disease. In the original paper Laveran described
and figured three different forms of the parasite: (1) pigmented, crescentic forms; (2) ameboid, flagellated organisms;
(3) spherical, non-motile bodies somewhat larger than the
red blood cell. In a subsequent paper he added to this (4)
spherical, highly refractile bodies smaller than the red blood
cell, and, as he believed, attached to the blood cell. He
believed these four varieties to be phases in the development
of a single parasite, and, supposing the flagellate forms most
important, proposed the name "*oscillaria malariæ.*"

Marchiafava and Celli (1885) confirmed and extended the
observations of Laveran. They observed segmenting forms,
and showed that the small bodies described by Laveran were
intracellular and not merely attached to the corpuscle. For
the small, non-pigmented form, to which they attached great
importance, they proposed the name "*plasmodium malariæ.*"

Golgi, whose researches upon the parasite of quartan fever
were published in 1885, took the view that the various forms
of parasite described by Laveran, Marchiafava, and Celli and
their followers represented, not a single polymorphous organism, but that there were three distinct types of organism,
each associated with a particular type of fever. The views
of Golgi are now generally accepted, and it may be said that
there is a distinct type of parasite associated with each clinical variety of malarial fever.

Types of the Malarial Organism.—(*a*) *Tertian type.*—
At the earliest stage of its cycle it is seen as a small, hyaline,

non-pigmented, intracellular body, which gradually increases in size, and, appropriating pigment, reaches full development in forty-eight hours. It undergoes segmentation at the time of the febrile paroxysm. The spores vary in number from fifteen to twenty. The young intracellular form is highly motile, and its outline is indistinct. The infected red blood cell is larger than normal, and rapidly becomes decolorized. This organism is associated with the tertian type of malarial fever.

(*b*) *Quartan type.* — The cycle of development requires seventy-two hours. The youngest form is a small, hyaline, endoglobular body resembling that of the tertian type. Ameboid movement is not so active. The organism is fully developed on the day of the fever, and is seen to fill the red blood cell, which is almost or entirely destroyed. Pigment accumulates in the center of the organism, and division into eight to ten pear-shaped bodies occurs. Segmentation begins eight to ten hours before the paroxysm.

(*c*) *Æstivo-autumnal type.*—This type of organism is associated with malarial fevers having an irregular course. The various phases in its cycle of development and the length of time required have not been thoroughly worked out as in the tertian and quartan types. Segmentation and much of the cycle of development goes on in the internal organs. It is believed that the length of the cycle varies, but that here, as in other types, the paroxysm is associated with segmentation and invasion of the blood by a new generation. These young forms are very scanty in the peripheral blood during the paroxysm, but can be obtained from the spleen and other internal organs. The segmenting forms are very scanty in the blood of all cases except those which are very severe. The manner of division in the æstivo-autumnal organism resembles that seen in the tertian type. After the fever associated with this type of organism has lasted a few days certain peculiar forms of the parasite appear in the blood—

the crescentic forms—which develop from large, ovoid, endo-globular bodies. Crescents are present in afebrile states; and it is believed that in this stage the organism does not give rise to marked symptoms. Crescents are seen in cachectic or anemic individuals. They resist the action of quinine, to which all other forms speedily succumb. Several theories have been put forward to explain their nature : they have been considered encysted forms of active, virulent parasites; as degenerated forms; they have also been supposed to represent the form in which the organism exists outside the body.

Marchiafava and Bignami have described two different varieties of the æstivo-autumnal parasite, one of which undergoes its cycle of development in twenty-four hours and the other in forty-eight hours and is an organism distinct from the tertian type above described. It may at present be said that while there may exist an organism which causes a quotidian fever and completes its cycle in twenty-four hours, other observers have not been able to follow it.

Flagellate Bodies.—The flagellate bodies described by Laveran appear in the cycle of all varieties of the malarial parasite. They are actively motile, and contain a central group of pigment granules. The filaments arise from different parts of the body, and are from two to three times the length of the body and have somewhat rounded ends. Flagellate forms are developed from the full-grown extra-cellular parasite. They are more abundant in blood from the internal organs than in the peripheral circulation. Quinine has no effect upon them. It has been supposed that the flagellate forms represent the first stage in the existence of the parasite outside the body; by others they are looked upon as degenerate forms. The exact significance of these bodies as regards the cycle of the parasite is not understood.

Classification.—The malarial parasite is by most authors placed in the class of Sporozoa.

Methods of Examination. — The most convenient, as well as the best, way of examining for the malarial parasite is in the fresh blood. For this purpose a small drop of blood is obtained by pricking the ear, and touched with a warm slide. Over this a cover-glass is gently laid, and the specimen examined with an oil-immersion lens. Dried blood specimens are fixed for five minutes in absolute alcohol and ether, equal parts. They are best stained according to Czenzynski's method. The parasite appears blue, and is always round in the stained specimen. The blood cell appears red.

Pathogenesis. — *Lower animals* have not been known to be infected with the malarial parasite.

Its relation to the disease in *man* can not be disputed. Its constant presence in the blood of malarial subjects, its absence in individuals in health and in other fevers, the regular exacerbation following sporulation, and its reaction to quinine all speak for it. More conclusive still are experimental injections. Subcutaneous or intravenous injections of the blood of a malarial individual will usually produce the disease in the second subject, with the appearance in his blood of an organism of the same type as that introduced. The average period of incubation after these inoculations is eleven or twelve days. Here, as in other diseases, one individual is more susceptible to the poison than another.

Little is known of the manner in which the plasmodium produces malaria. It is the supposition that a poison originates in the cell remnant, which breaks down after giving off spores. Some have tried to explain the action by a mere mechanical presence of the parasites in the blood.

The manner in which the parasites invade the body is at present a subject of doubt. While most authors have looked upon the lung as the avenue by which they enter, later authors have attached importance to the sting of insects (mosquitoes) in carrying the disease-producing cells.

The incubation period varies from six to ten days to several months.

Immunity Experiments.—Recovery from an attack of malaria seems rather to favor its recurrence than to cause immunity. All efforts at immunization have been futile.

TEXAS OR SPLENETIC FEVER.

The disease known as Texas or splenetic fever is a specific affection of cattle, due to invasion of the blood by a hematozoön which gains entrance through the body of the cattle tick, *boöphilus bovis.* It is characterized clinically, in acute cases, by high fever, rapid disintegration of the red blood-corpuscles, and hemoglobinuria. In the chronic forms there is little or no fever; hemoglobinuria is not observed. The principal anatomical changes after death are found in the liver and spleen.

The Micro-organism of Texas Fever.—In the blood of acute cases during life the parasite occurs in pairs (sometimes singly) within the red blood cell; rarely it is found free. It is pyriform in outline, devoid of pigment, homogeneous, from 2 μ to 4 μ in length, and from 1.5 μ to 2 μ in width at the widest portion. Five or six hours after death these pear-shaped bodies are seen to have become round. The inference is that they have assumed this shape under the adverse conditions brought about by the death of the host.

The mild or autumnal type of Texas fever is characterized by the presence of an endoglobular parasite, rounded in form, and from 0.2 μ to 0.5 μ in diameter. From five to fifty per cent of the red blood cells may be infected with this form of the parasite for a period of two to five weeks. (Smith.)

Staining.—The parasite stains with hematoxylon and methyl violet, but best with Loeffler's alkaline methylene blue. It is often difficult to demonstrate it in the blood during life. They may be few or entirely absent when the number of corpuscles have sunk very low, yet in such cases infected corpuscles will be very numerous in areas of stasis.

Transmission of Texas Fever.—The cattle tick, *boöphilus bovis* (Curtice), infests cattle in the southern part of the United States, and is the carrier of the microparasite from the immune animals of this region to susceptible ones of the North. Cattle from southern points carry on their bodies during the summer months large numbers of these infective ticks. The females, when matured, drop to the ground, where the process of laying begins. An adult female tick will lay about 2,500 eggs. These hatch, and the young ticks, finding their way upon susceptible cattle, induce in them Texas fever. Disease does not follow the introduction of infective cattle into a non-infected territory until a new generation of ticks appear.

There is no certainty as to whether the endoglobular parasite of Texas fever belongs primarily to southern cattle, and is drawn from their blood by the tick, or whether it is primarily of the tick itself. It can not be found by microscopical examination of the blood of tick-bearing southern cattle, yet injections of blood of these animals, freed of ticks for three or four months, give rise to Texas fever. The micro-organism is resident most probably in small numbers in the blood of immune southern cattle, and the tick acts merely as a carrier of infection.

The cattle tick sometimes infests horses, dogs, and other animals, but they are not affected by its presence. Attempts to induce the disease in other animals have failed.

Pathogenesis.—A period of eight or ten days elapses after the young tick has attached itself before the appearance of symptoms. How the microparasite invades the blood is not known. Australian observers claim to have discovered it in the body of the tick. We are equally uncertain as to its life history within the body of affected cattle. It is certain, however, that it there undergoes very rapid multiplication, that it invades and destroys the red blood cell.

Immunity Experiments.—One attack of Texas fever confers immunity against a subsequent fatal attack. Calves are immune. It is said that cattle bred in areas of permanent infection lose their immunity after three years' removal. Non-immune cattle taken into infected districts are attacked almost without exception.

THE BACTERIOLOGICAL ANALYSIS OF WATER, AIR, AND SOIL.

In determining the source of the various infectious diseases and the manner of their dissemination, the bacteriological examination of the water we drink, of the air we inhale, and of the soil upon which we live is of importance sufficient to justify special consideration. These examinations have gained in importance in the last years, especially the bacteriological study of water intended for domestic use.

Water.—It is known to-day that polluted water of rivers and wells often produces disease, hence the importance of a knowledge of the methods employed in investigating the bacteria which may be present.

All waters in their natural state contain bacteria. The great number of them are non-pathogenic—the so-called "water bacteria." They overgrow pathogenic species which may be present, acting in this way as disinfectants. The pathogenic varieties of greatest importance found in water are the vibrio of cholera and the typhoid bacillus. The colon bacillus and the staphylococci are also frequent inhabitants of polluted water. Any of the pathogenic species may exceptionally be found. The number of bacteria present in water is dependent upon the character and extent of the pollution and the season of the year. During the summer months the number diminishes near the surface of the water, owing to the bactericidal power of sunlight and oxygen. While this purification is, in consequence of superficial action, only slight in standing or slowly running water, it attains greater proportion in rapidly-flowing streams, where new portions of the water reach the surface to become exposed to the sun's rays.

The number and variety of species also varies in different waters. It has been shown, for instance, that spring water contains 2 to 50 bacteria, while river water contains 1,000 to 20,000 in each c.cm. Their isolation is of little importance as compared to the detection of disease-producing species. The number of the pathogenic forms bears an inverse relation to the number of saprophytic species present. Their presence in the water is due to contamination with excreta of infected individuals. The proportionally small number in which pathogenic varieties are present makes their isolation difficult and entails much labor.

In the technique of examining water rigid precautions are necessary to prevent the entrance of germs other than those originally present. The water is usually obtained from under the surface in sterile flasks. These are plugged with sterile cotton. The examination should be undertaken at once, as in a very short time considerable change takes place in the relative number of the different species present as well as in their total number.

Should an immediate examination not be possible, the specimens must be kept on ice in order to prevent, as much as possible, bacterial growth.

After shaking the water, to distribute the germs equally, one or two drops are taken from it with a sterile loop, and stirred into liquefied gelatin. This is poured upon plates and allowed to congeal. After some time colonies appear, which can be transferred to other tubes to study their characteristics in pure culture.

As the pathogenic germs are present in much smaller quantities than the non-pathogenic, their detection is difficult. This difficulty, and the fact that outside of the cholera vibrio and the typhoid bacillus the water seldom contains pathogenic species, has led to the introducing of special methods for the isolation of these two varieties.

Detection of the Cholera Vibrio.—If the presence of the cholera vibrio is suspected in water, nutrient substances are added to favor its growth. The water is then allowed to stand for a given time at a temperature (37° C.) favorable for the development of the cholera germ, and less so for the "water bacteria," which brings about a relative increase of the former. The chemicals added to the water differ in the methods of different authors. Heim added two per cent of peptone and one per cent chloride of sodium; Flügge made a one-per-cent peptone solution in water, and Koch added one per cent peptone and one per cent chloride of sodium. They allowed the water to set at 37° C., and after ten, fifteen, and twenty hours transferred particles of the growth which had formed on the surface to tubes.

If the cholera vibrio is not found by this method, we can not absolutely say that the water does not contain it; for (1) it may not have been included in the small volume of water used in the test, and (2) it may have been overgrown by other species.

Detection of the Typhoid Bacillus. — Bacteriological examination of water for the typhoid bacillus is of much importance, since it has been shown that nearly all cases of typhoid fever infection arise from polluted water. In these examinations difficulties are encountered by reason of the small numbers in which the bacillus is present relative to other organisms, and from the absence of any characteristic by which the typhoid bacillus can be readily distinguished from numerous forms which resemble it.

Chantemesse and Widal, in 1887, noting that the typhoid bacillus grows in weak solutions of carbolic acid, suggested the addition of this agent in the proportion of 0.2 per cent to nutrient gelatin. Carbolic acid in this proportion, however, has a decided inhibiting power upon the typhoid bacillus, and does not restrain to an equal extent many other species.

Holz, in 1890, prepared a gelatin medium from the juice of raw potatoes. (See culture media.) Upon this many saprophytic species will not develop. To increase this inhibiting power 0.2 per cent carbolic acid was added.

Elsner (1895) improved the medium of Holz by substituting for the carbolic acid 0.75 to 1 per cent potassium iodide. He recommends that the gelatin be made of such a degree of acidity that 10 c.cm. will be neutralized by from 2.5 to 3 c.cm. of a decinormal solution of caustic soda.

Capaldi's medium for isolating the typhoid bacillus has the following composition :

Water ....1,000 c.cm.	
Peptone................................	20 grams.
Gelatin.................................	10 "
Grape sugar	2 "
Sodium chloride.......................	5 "
Potassium chloride....................	5 "

This is boiled, filtered, and made alkaline, after which 2 per cent of agar is dissolved in it. The medium is filtered and sterilized. The advantage of this medium is that it can be used for cultivation at incubator temperature. We have found, however, that the addition of 1 per cent potassium iodide to nutrient agar answers equally as well. Both allow other species to develop. Upon the media of Holz, Elsner, Capaldi, and the iodized agar the typhoid bacillus and bacillus coli communis are the predominating species which develop. Upon the gelatin media colonies appear on the third or fourth day. The typhoid colony is small, whitish, pin-point sized, while that of the bacillus coli communis is larger, and under the microscope is round or oval, of a brownish color and granular texture. Development is more rapid in the agar media, typhoid colonies being visible in eighteen to twenty-four hours. For further study colonies resembling the typhoid bacillus may be transferred to sterile tubes of nutrient gelatin and agar.

Liquid media are of service for the reason that larger quantities of suspected water may be used with them. Parietti's fluid has the following composition:

Acid carbolic........................... 5 grams.

Acid hydrochloride..................... 4 "

Distilled water ....100 "

To 10 c.cm. sterile neutral bouillon is added from one to ten drops of this fluid. They are then placed in the incubator for twenty-four hours, and to those tubes which remain clear at the end of this time five to ten drops of the suspected water is added. At the end of twenty-four hours the medium becomes cloudy. Gelatin or agar plates are now made. Parietti's fluid suppresses many, but not all, saprophytic species.

Gruber and Durham (1894) took advantage of the specific paralyzing action of typhoid serum upon the bacillus of Eberth. They added to neutral bouillon tubes eighteen to twenty-four hours after inoculation with the suspected water the blood-serum of an animal immunized against the typhoid bacillus. As the result of the addition of this serum the typhoid bacillus, if present, after five or six hours falls to the bottom of the tube.

The specific action of typhoid serum has shortened the process of isolating the typhoid bacillus from water. The method used by us is the following: A few drops of the suspected water is added to bouillon tubes containing Parietti's solution and placed in the incubator. As soon as cloudiness appears in the bouillon, agar plates are made and kept at 37° C. In twenty-four to forty-eight hours the typhoid colonies can be detected. These are fished out and the action of typhoid serum tested upon them. Reaction to this test is sufficient to enable one to say that the organism is the typhoid bacillus. Wyatt Johnston has recently suggested that the suspected water be added to neutral bouillon and incubated for twenty-four hours. At the end of this time it is sterilized and injected into rabbits. The reaction of agglutination appears in five or six days if the typhoid bacillus was present.

Quantitative Bacteriological Analysis of Water.—In Koch's original method of making a quantitative test of the bacteria contained in water, a certain quantity ($\frac{1}{2}$ or 1 c.cm.) was stirred into gelatin, which was poured on a plate and allowed to congeal. If the germs were present in large quantities, the fluid was diluted with sterile water in given proportions (ten to twenty parts).

When the colonies developed they were counted by a special apparatus devised by Wolffhuegel. It consists of a horizontal black plate upon which the plate of cultures is laid. Close above the gelatin a clear glass rests. It is delicately marked off in regular squares. (These may be marked on the black background instead.) The colonies in a number of squares are counted with the aid of a lens. By multiplying the average number in each square by the number of squares on the plate, the number of colonies on the plate is arrived at.

If the colonies are densely crowded, a better method of counting them is with the aid of the microscope (low power). The colonies in a number of fields (twenty or more) are counted to find the average number in each field. Knowing the ocular aperture in use and the number of colonies in a field, the number upon the entire plate can readily be estimated.

If the colonies are still too dense to count, a method of diminishing the ocular aperture in known proportion by diaphragms was suggested by Ehrlich.

In a Von Esmarch tube colonies can be counted by dividing a circular area of the tube into squares of equal size.

These methods give a reasonably accurate estimate of the number of bacteria contained in the amount of water examined. A number of bacteria thrown together may go to form the same colony, and in this way reduce the number some.

Note.—The methods given for the examination of water can be applied to other fluids as well.

Air.—Bacteriological analysis of air came into prominence through the study of wound infection. Bacteria in the air are readily accounted for in a drying-up, after their development elsewhere, and being picked up by currents of air.

Bacteria found in the air are mostly saprophytes, although pathogenic varieties—particularly the pyogenic cocci and tubercle bacillus—are known to exist.

The air of our daily surroundings, especially of cities, contains most bacteria (100 to 1,000 in one c.cm.), while out at sea and on mountain peaks the air is almost free from germ life. The species contained are usually micrococci. Sarcina are also nearly always present. Chromogenic species are not uncommon. But few have the property of liquefying gelatin.

To determine the species contained, a sterile gelatin plate is allowed to set in the air to be examined for a while. The plate is put away and colonies allowed to develop.

A better method, and one which answers for qualitative as well as quantitative analysis, is that suggested by Petri. By means of a gauged air-pump he forced a certain amount (50 to 100 liters) of air through fine sterilized sand packed in small glass cylinders (8 c.cm. + 1.51 to 1.8 c.cm.). The sand which retains the bacteria is mixed into a tube of liquefied gelatin.

Plates are poured and colonies allowed to develop. For quantitative test the colonies can be counted as in the examination of water.

As it is not always easy to differentiate between the grains of sand and colonies on the plates, Sedgewick and Tucker have substituted sugar for the sand.

Other methods, such as passing air through water and examining this bacteriologically, or of passing it through liquefied gelatin and pouring plates of this, are inferior to Petri's method.

These means have all failed to isolate the pathogenic species. They have, in fact, never been isolated from the air direct, though animal experiments and all other known means have been applied. This can be accounted for in the relatively small number of pathogenic species present. It has, however, indirectly been demonstrated that they exist in the air. Carnet, who worked particularly with the tubercle bacillus, produced tuberculosis in animals by introducing (on a moist sponge) dust taken from a place which had a short while previously been thoroughly sterilized. The presence of bacilli on the previously sterile place could be accounted for only in their gravitation from the air.

Soil.—The bacteriological analysis of soil has no great value as far as pathogenic species are concerned. It, however, merits mention for the light it has thrown upon the process of decomposition of dead animal tissue. The soil near the surface contains myriads of micro-organisms, principally bacilli; 1 c.cm. was shown to contain 100,000 bacteria. They gradually diminish in number and almost cease to be present at $1\frac{1}{4}$ meters. The bacilli have marked property of decomposing organic tissue, which, by being split up into simpler elements, is prepared to furnish nutrition for growing plants.

Of the pathogenic species, the bacillus of malignant edema is most frequently found in the soil. The tetanus bacillus and the bacillus of symptomatic anthrax are also frequent.

In the technique of examining the soil bacteriologically small quantities of earth are secured by means of a special instrument introduced by C. Fraenkel. By means of the instrument earth can be taken from any desirable depth without contamination in its course through the passage made by it. A certain quantity of the earth is broken up and mixed into liquid gelatin, which is allowed to congeal on the walls of the tube (Von Esmarch). Earth should, just as well as water, be examined as soon as possible after it is secured.

The Von Esmarch tube is preferable to plates because it is impossible to pour the gelatin out of the tube without leaving back numerous grains of earth and consequently bacteria.

To pour plates the earth is broken up in sterile water, and the plates made from the water in the customary manner.

The presence of the pathogenic varieties is best demonstrated by injecting small portions of the suspected earth into lower animals (mice). If present in but small quantities the animal develops the corresponding disease and dies. Bacteria can then be cultivated from the dead body.

www.ingramcontent.com/pod-product-compliance
Lightning Source LLC
Chambersburg PA
CBHW021709210326
41599CB00013B/1585